Savoring Alternative Food

Advocates of the alternative food movement often insist that food is our "common ground"—that through the very basic human need to eat, we all become entwined in a network of mutual solidarity. In this challenging book, the author explores the contradictions and shortcomings of alternative food activism by examining specific endeavors of the movement through various lenses of social difference—including class, race, gender, and age.

While the solidarity adage has inspired many, it is shown that this has also had the unfortunate effect of promoting sameness over difference, eschewing inequities in an effort to focus on being "together at the table." The author explores questions of who belongs at the table of alternative food, and who gets to decide what is eaten there; and what is at stake when alternative food practices become the model for what is right to eat? Case studies are presented based on fieldwork in two distinct loci of alternative food organizing: school gardens and Slow Food movements in Berkeley, California and rural Nova Scotia. The stories take social difference as a starting point, but they also focus specifically on the complexities of sensory experience—how material bodies take up social difference, both confirming and disrupting it, in the visceral processes of eating.

Overall the book demonstrates the importance of moving beyond a promotion of universal "shoulds" of eating, and towards a practice of food activism that is more sensitive to issues of social and material difference.

Jessica Hayes-Conroy is Assistant Professor of Women's Studies, Hobart and William Smith Colleges, Geneva, New York, USA.

Routledge Studies in Food, Society and Environment

Street Food
Culture, economy, health and governance
Edited by Ryzia De Cássia Vieira Cardoso, Michèle Companion, Stefano Roberto Marras

Savoring Alternative Food
School gardens, healthy eating and visceral difference
Jessica Hayes-Conroy

Food Sovereignty in International Context
Discourse, politics and practice of place
Edited by Amy Trauger

For further details please visit the series page on the Routledge website:
www.routledge.com/books/series/RSFSE/

Savoring Alternative Food

School gardens, healthy eating and visceral difference

Jessica Hayes-Conroy

Routledge
Taylor & Francis Group

LONDON AND NEW YORK

First published 2014 by Routledge

2 Park Square, Milton Park, Abingdon, Oxfordshire OX14 4RN
711 Third Avenue, New York, NY 10017

Routledge is an imprint of the Taylor & Francis Group, an informa business

First issued in paperback 2017

British Library Cataloguing-in-Publication Data
A catalogue record for this book is available from the British Library

Library of Congress Cataloging-in-Publication Data
Hayes-Conroy, Jessica.
  Savoring alternative food: school gardens, healthy eating and visceral
  difference / Jessica Hayes-Conroy.
  pages cm.—(Routledge studies in food, society and environment)
  1. Food habits—Social aspects. 2. Nutrition—Social aspects. 3. Local
  foods—Social aspects. 4. Sustainable agriculture—Social aspects.
  I. Title.
  GT2860.H39 2014
  394.1′2—dc23
  2014014751

ISBN: 978-0-415-84423-9 (hbk)
ISBN: 978-0-8153-9533-1 (pbk)

Typeset in Bembo
by Florence Production Ltd, Stoodleigh, Devon, UK

To my twin sister, best friend, and closest colleague, Allison Hayes-Conroy.

Contents

Acknowledgments

There are many people that I should acknowledge for their help in supporting me as I conducted the research and writing for this book. Most importantly, I would like to thank James McCarthy, Julie Guthman, Kai Schafft, Lorraine Dowler, and Sandra Morgen for their hard work in reading and commenting on earlier drafts of this research project. I would also like to acknowledge the kindness and generosity of so many of my research participants, in both case study locations. I would not have had the opportunities for learning that I was afforded in this research were it not for the help of these kind and generous people, some of whom have become dear friends. I also would not have been able to conduct this research without the support of the Society of Women Geographers Fellowship Award, as well as the Graduate Student Fellowship Program Grant from the Canadian Embassy.

A number of friends and colleagues supported me during this work, among them Sasha Davis, Dana Cuomo, Allyson Jacobs, Matt Jacobs, Alissa Rossi, Jairus Rossi, Jason Beery, Melissa Rock, Thomas Sigler, Shaunna Barnhart, Rafael Díaz-Torres, Hector Saez, Kerry Gagnon, Robin Lewis, Kristine Johanson, and Kendralin Freeman. Finally, I could not have completed the work without my sister Allison Hayes-Conroy, to whom this book is dedicated, and also without the support of the rest of my family, including Linda Hayes and Rusty Conroy, Jennifer Douglass, Victor Gutierrez Velez, and Benjamin Velez-Hayes.

Abbreviations

ANT   Actor Network Theory
ESY   Edible School Yard
NRT   Non-Representational Theory
SGCP   School Garden and Cooking Program
USDA   United States Department of Agriculture

Introduction

Eating sea urchin for breakfast

"Look at your faces!" Peter exclaimed. "You're answering your own questions." I looked at Allison, wondering whether my own face gleamed with surprised curiosity as hers did, or if it betrayed the underlying nausea that involuntarily rose up within me. It was eight in the morning, after all, and the spoonful of bright orange, raw sea urchin (the reproductive organs of the animal) just didn't seem to mix well with the French pressed coffee that I had just gulped down. I pondered the description he gave us: "It's like a sea pudding fresh from the ocean." But there was little time to dwell on this; a minute later we were climbing the stairs to the restaurant's walk-in fridge for a sample of Peter's homemade lardo, followed by an earnest admiration of his just-picked chioggia beets. My awe was doubled, I admit, by the fact that I secretly wished we could have tried those instead. But Peter was one of the most highly respected chefs in Nova Scotia, and wonderfully friendly at that. And so I chided my wishful thoughts, and despised my underdeveloped palate for not being able to appreciate his generosity to the fullest extent.

Our meeting with Peter that humid, bright morning in mid-July was both commonplace and peculiar, mundane and exceptional. It was a typical moment, on a typical day, of a six-month research trip that was, if one word could describe it, extraordinary. This book tells the stories that I experienced and accumulated during that trip. It explores the lessons that I (along with my closest colleague, who also happens to be my twin sister) gleaned from the others we met who were engaged in the world of 'alternative food' activism—that is, the promotion of local, sustainable, and healthy food systems. But unlike a lot of popular books on dieting and food, quick to tell the reader what to eat and why to eat it, my stories address the complexities of taste and eating, and the contradictions that compel multiple, even contradictory reactions towards food. My research taught me not about the simplicity of good consumption, but about the complicated ways that food, bodies, and social identities intersect. Thus, the purpose of my storytelling is not to leave the reader with a neat distillation of the 'shoulds' of eating. Rather, what I hope to do in the pages that follow is to promote a more expansive, humble, and sensuous approach to food—one that is soaked in real-life contradictions, steeped in social difference, and dripping with both desire and disgust.

Leaders of the alternative food movement often insist that food is our common ground, and that through the very basic human need to eat, we all become entwined in a network of mutual solidarity. This oft-repeated adage, while inspiring to many, has also had the unfortunate effect of encouraging alternative food activists to overlook other stories: the stories of those who are not invited to the table, either through lack of access, cultural familiarity, or personal desire. In this introductory chapter, I discuss such complexities and contradictions of alternative food activism, focusing on the region of North America, and describing in depth the work of two School Garden and Cooking Programs (SGCPs), as well as the alternative food activism that surrounds them. Using anecdotes from my own work in California and Nova Scotia, I explore the alternative food movement's goals and strategies, as well as its historical trajectory, and I discuss how and why alternative food activism has also come under some serious critique. I also highlight key players in this movement—from the ongoing social outreach work of famous chefs like Alice Waters and Jamie Oliver to the political work involved in Michelle Obama's recent healthy eating campaign—and I use narrative accounts from conversations with students, parents, and teachers to illustrate the varied public reaction to such initiatives. Ultimately, the goal of this introduction is to demonstrate, through example, the importance of moving beyond a discussion of the "shoulds" of eating—that is, of specifying what is right or wrong to ingest. Instead, this introduction serves up something quite different: a motivation to embrace food in all its complexities and contradictions—an invitation to savor the uncommon.

Place and bodies in food research

To conduct this research, my twin sister, Allison, and I spent 3 months each in two seemingly very different places: rural Nova Scotia and Berkeley, California. As geographers and social scientists by training, we wanted to explore how alternative food activism plays out in different geographic and social contexts. Berkeley, of course, was an obvious choice for a number of reasons. Arguably the North American capital for alternative food activism, Berkeley hosts numerous leaders, thinkers, and practitioners of alternative food, including big names like Alice Waters, restaurants like Chez Panisse, and ideas like The Edible Schoolyard. Rural Nova Scotia is, in many ways, Berkeley's counterpoint. A rural hinterland to the comparatively small urban center of Halifax, Nova Scotia's rural communities boast neither big names nor leading ideas in the movement towards alternative food. And yet—perhaps standing as a testament to the ubiquity of food concerns more broadly in North America—alternative food activism is alive and well in Nova Scotia. Indeed, the alternative ideas and food practices so familiar to residents of Berkeley have also found their way to the dinner tables and community schoolyards of many Nova Scotians.

Still, rural Nova Scotia and Berkeley, California are distinct in several important ways. For one, Berkeley is much more diverse than rural Nova Scotia in ethnic and racial terms, while the key dividing line among most rural Nova Scotians tends to be social class. Berkeley also possesses a cosmopolitan cachet to which rural Nova Scotia can only (and does) aspire. And the geographic realities of the Bay Area of California make alternative food practices more easily realizable in Berkeley, in terms of both climate and critical mass. These distinctions intrigued me, especially because I was acutely aware, as I began my work, that the need for alternative food systems had not (and still *has* not) been universally accepted across the social landscape. Indeed, both academics and activists have roundly critiqued the alternative food movement for its racial homogeneity and elitist bent—essentially for being largely an upper-middle-class, white movement. It seemed to me, then, that in order to really understand how alternative food activism encounters and engages social difference, I needed to pay close attention to how the practices of alternative food play out in particular locations; that is, how and why the movement spreads, attracts new participants, and gains momentum (or doesn't, as the case may be). Ultimately, I was interested in answering *not* the question of why people should, or shouldn't, engage in alternative food activism, but instead the question of why they did, or didn't. In other words, I was interested in the question of motivation.

As researchers, to be motivated by motivation means to be propelled by questions of and for the body, in addition to the intellectual mind. It means that we must pay attention to the sensations and feelings that we experience through our own bodies, and also that we must strive to attend to the ways that other bodies are similarly or differently affected. To focus on bodily motivation, however, does not mean dismissing other questions of social identity, position, and difference. On the contrary, it means that we must become attentive to the ways that sensory perceptions and visceral reactions both propel and betray social categories—for example, the way that our race, class, gender, and age comes to influence, and be influenced by, what we desire and detest on our dinner plates. It means that we must struggle to understand how social disenfranchisement and physical tastes both reinforce and resist each other—as when limited access to a grocery store leads to a taste for fast food, and thus a higher risk for diabetes and other dietary diseases, furthering one's economic and social hardship. To be sure, these are age-old questions in the social sciences, questions that link social structure to individual agency, and societal pressures to personal drive. Certainly Pierre Bourdieu's now classic work on social class and the judgment of taste stands as an important example of this kind of academic inquiry (Bourdieu 1984). And yet, to attend to one's body as *itself* an active agent in the promotion of social change is a livelier task than what has typically been attempted in the social sciences. It is an intellectual task, in part, but it is also one that is infused with a liveliness and a hopefulness of the palate— an openness that comes from the recognition that we all have the capacity to create and compel social change through our taste buds, and thus to resist and recreate both who we are, and who we are becoming.

It is for this reason that the stories I tell in this book are stories written about, and with the help of, a variety of different bodies. Allison and I came to recognize over the course of our travels that the physical sensations, moods, and attachments of people's bodies are not secondary to the work of alternative food activism. In fact, as with much social movement organizing, the success of alternative food as a force of social change depends largely upon the movement's ability to motivate bodies to engage in alternative practices—to generate sensations of joy, pleasure, and belonging, rather than hostility, aversion, and dislocation. The fact that this currently happens within some bodies and not others is a key, I argue, to understanding the homogenizing and alienating feel that alternative food can sometimes provoke. Therefore, paying attention to the role of the material body in different people's engagement with alternative food can be useful in helping to understand how issues of social difference intersect with alternative food activism in distinct geographic and social contexts.

To narrow and direct our focus in undertaking this work, Allison and I decided to explore two particular alternative food initiatives, both of which reveal a decisive focus on sensory perception and the material body. Allison's research focused on the Slow Food movement, an international organization with local chapters or 'convivia' throughout much of Europe and North America. Slow Food is a movement strongly dedicated to the physical senses, with efforts in taste education and taste preservation. Many of Slow Food's local convivia are strongly invested in activities that address the bodily experience of eating, as much if not more than any intellectual or conceptual understanding of food. I was lucky to be able to accompany Allison on many of her Slow Food research trips, including interviews and participant observation. Through this work, I came to understand much about the terrain of alternative food (both Slow Food and other initiatives) in the areas and peoples that surrounded my own research focus.

Often related conceptually if not also financially to the Slow Food movement, School Garden and Cooking Programs (SGCPs) promote hands-on educational experiences that are again grounded in sensory education, as opposed to cognitive learning. SGCPs engage students in direct, body-centered activities both in garden and kitchen classrooms; students dig, plant, weed, and harvest a bounty of fresh produce, and then haul it into kitchen classrooms where they learn to chop, sauté, season, and bake their harvest into a variety of seasonal snacks. Focusing on day-to-day life within SGCPs gave me the opportunity to explore the role of the physical, material body in specific instances of alternative food activism, and it allowed me to quite easily broach the subject of bodily motivation with those who were engaged in hands-on, food-based activities. More importantly, while my connection to Allison's Slow Food research allowed me to better understand a well-organized, adult-centered, and socially homogeneous alternative food initiative, my focus on SGCPs provided an opportunity to explore people's experiences of alternative food activism in a much more socially diverse and largely kid-centered context.

One more note on this research is necessary here. The fact that Allison and I are identical twins is not, I believe, incidental to the stories that we are able to relate. To the contrary, it was precisely because of the uniqueness of our position as researchers—the extent to which we look and sound alike, the physical obviousness with which we display our familial connection, and the rareness of our intellectual bond—that our research unfolded as it did. Indeed, as we repeatedly experienced, people opened their hearts to us (and often their homes) not in the collegial way in which a stranger might humor an academic's request for more information, but with a mixture of intrigue and care, as if the evident facts of our kindred identities—as sisters to each other, and daughters to distant others—made our questions all the more compelling and vital. Furthermore, having a research and travel companion with an intellectual mind and physical body so closely related to one's own allowed us both the opportunity to deeply interrogate our own visceral experiences—to collectively discuss how and why they converged or conflicted with each other, and when and where they confirmed or denied our social and biological likeness. In this sense, our ability to consciously observe and experience our own bodies throughout this research process only made us more aware of the contradictions and complexities of body-centered work, and also made us more open to the possibilities and chance occurrences that such bodily interactions can engender.

Studying healthy alternatives

> I am not building this as a model for the city, or for the nation; I am building it as a model for the world.
>
> (Activist, CA, paraphrasing Alice Waters on the purpose of The Edible Schoolyard at its inception)

On November 10, 2008, *The New York Times* ran an article entitled "Bake Sales Fall Victim to Push for Healthier Foods," about how school nutrition rules are impacting school traditions (Brown 2008). The article discussed nationwide trends in school food, but focused particularly on the changes that are being implemented in California schools, including the Berkeley school district. The write-up contained a photograph of a kindergarten teacher reviewing a list of "good foods" and "bad foods" with her five-year-old students. The list of bad foods included candy, chocolate, Burger King, McDonalds, cake, ice cream, and soda. Salad and strawberries topped the good food list.

In March of 2009, First Lady Michelle Obama announced her plans to dig up part of the White House lawn in order to make space for an organic vegetable garden. The impetus for this garden was both ecological sustainability—reducing greenhouse gas emissions that arise from industrial production and distribution—and human health—encouraging fresh fruit and vegetable consumption "at a time when obesity and diabetes have become a national concern" (Burros, March 19, 2009). The garden is meant to become an inspiration particularly for children. "My hope," said Mrs. Obama, "is that through children, they will begin to educate their families and that will, in turn, begin to educate our communities" (Burros, March 19, 2009).

Both of these examples are representative of a growing emphasis in the United States and Canada on what is broadly called "alternative food"—a term that has come to signify a variety of food practices that are intended to counter the perceived ill effects of the modern/industrial food industry on both ecological and human health. From its origins in the environmental movement of the 1960s (among other origins), the alternative food movement has emerged as a critique of industrial methods of agricultural production and distribution (from Rachel

Carson's famous exposé of the dangerous insectiside DDT in 1962 to Vandana Shiva's attack on genetically modified foods in 2000; see Carson 2002, Shiva 2000). The push for alternative food has also involved an increased emphasis on local growing and local purchasing, such that the social relations of production and consumption can (ideally) be conducted at a fully knowable, 'human' scale (Hinrichs and Lyson 2009, see also Born and Purcell 2006). This, in turn (it is argued), supports the economic and social well-being of rural communities, and keeps the food system out of the hands of large-scale agribusiness corporations, which tend to be inattentive to the needs of local communities and landscapes. More recently, with growing concerns over the so-called obesity epidemic, this push for alternative foods has also occurred alongside increased widespread attention to dieting, body size, and healthy eating (Pollan 2003, Nestle 2007). More and more alternative food initiatives now tend to fall at the intersection of environmentalism and nutrition, thereby linking ecological health and community health with the health of human bodies. SGCPs, including Alice Waters' now famous Edible Schoolyard project in Berkeley, CA, are prime examples of this growing trend (Waters 2007b, 2009).

SGCPs are alternative food initiatives that seek to encourage healthy eating habits in children by offering hands-on, sensory-based experiences in garden and kitchen "classrooms." The idea is that through these intimate, sustainable food experiences, children will come to choose healthier foods, including locally grown fresh fruits and vegetables. While neither school gardens nor cooking education are necessarily new phenomena in schools (think of Waldorf-inspired schools, or home economics classes), SGCPs have recently gained notoriety and momentum within North America under the converging contexts of ecological/economic sustainability and human health concerns, particularly because they are seen as projects that creatively address all of these issues simultaneously. What's more, SGCP leaders have sometimes been able to find considerable financial support for their alternative school food initiatives by framing them as also or especially healthy eating projects, instead of projects that simply attack the industrial food system. (The US Department of Agriculture, for example, which also supports agribusiness, funds school gardens as nutrition intervention projects through food-stamp related grant money; see USDA 2014).

This book explores this growing trend of healthy, alternative food in schools by way of two North American SGCP case studies: a public middle school in Berkeley, California (hereafter called Central School, a pseudonym used for the purpose of confidentiality), and a public K-6 school in a rural community in Nova Scotia (hereafter called Plainville School). Although neither of these projects were directly connected to The Edible Schoolyard (ESY) project of Alice Waters and the Chez Panisse Foundation (see edibleschoolyard.org and chezpanissefoundation. org), both projects have been influenced by not only the media attention and public support that the ESY project has garnered, but also by the ideological frame of "universal" importance that Alice Waters and other alternative food leaders have lent to the ESY project (i.e. as a "model for the world"). In other important ways, however, the case studies that I focus on differ dramatically from the ESY project.

Most notably, the two projects I studied rely on limited public grant money and much volunteer labor, arguably making them somewhat more realistic "models for the world" than the generously supported ESY project, which relies heavily on private funds.

The discussions herein draw from over 3 months of in-depth, qualitative research at each case study location, including interviews with teachers, parents, and leaders of the SGCPs, as well as focus groups and in-class activities with students, and many hours of participant observation in the kitchen and garden classrooms. As I already mentioned, in each location I also conducted interviews with a variety of alternative food leaders (including and beyond Slow Food activists) who were active more broadly in the community, beyond the functioning of the specific SGCPs in question. The purpose of these interviews was to try to understand what types of food-related struggles each community was facing, and how (if at all) the SGCPs fit into these larger food struggles. The data from all of this fieldwork was subsequently transcribed and coded thematically, and then analyzed in light of both the patterns/tendencies and exclusivities/potentialities that emerged from my transcripts and memos. Throughout the book, I present the findings of this qualitative work through the use of interview quotations, field note excerpts, anecdotes and detailed description.

Academically speaking, this book weaves together elements from three broad fields of intellectual scholarship: human-environment geography (particularly political ecology and agri-food studies), human/cultural geography (particularly geographies of affect and emotion), and women's studies (particularly theories of embodiment and social difference). In my theoretical argumentation, all three of these areas converge into what I have called "visceral geographies," which I have carved out as the academic space in which I seek to analyze the case study SGCPs (also see Hayes-Conroy and Hayes-Conroy 2008, 2010b). The topic of viscerality is the centerpiece of Chapter 1, and so I will refrain from a detailed discussion here; however, generally speaking visceral geographies refer to the ways in which bodily human experiences are reproduced *relationally* and *materially* through complex interconnections with both social forces and other physical 'bodies' (including food and food environments). In short, a visceral geography approach to the SGCPs imagines bodily motivation to eat certain foods as something that is variously and contextually produced through a wide array of social relationships, intellectual engagements, and material attachments, which give rise to explainable but *not* predetermined eating experiences.

This way of approaching the SGCPs makes sense for a variety of reasons, but most basically because it allows me to think through how food is at once a matter of social and biological relevance—following from food's connections to ecological sustainability, community well-being, and human health. The relational and material approach also allows me to be attentive to the power relations that are necessarily embedded within individual food behaviors, motivations, and 'choices,' while at the same time be open to uncovering how the SGCPs might also encourage changes in the power structures that configure both alternative and industrial food systems. In short, the concept of visceral geographies provides a lens through which to consider how the (uneven) social and material relations of our food system *come to matter* in the playing out of alternative food politics within human bodily experiences. Importantly, however, in doing such work I do not seek to argue that particular sets of relations can or will always produce the same bodily experiences or sensations. Rather, I draw on events and occurrences from my case studies to describe how bodily motivation to eat is at once structured and haphazard—how it is both dependent on social positioning and ever-open to new outcomes.

Awareness and difference in alternative food

Because I consider myself to be an alternative food advocate, and ultimately a proponent of SGCPs, it was especially important for me in conducting this research to be aware of the past and present limitations of the alternative food movement. More particularly, I wanted to be cognizant of the ways in which alternative food activism can both counter and reinforce social hierarchies. As SGCPs have emerged within the broader rubric of healthy food alternatives, it is clear that they have also taken on many of the problems and contradictions of such food-based activism. One especially telling result has been the framing of SGCPs, like so many other alternative food projects, as initiatives that ultimately seek to counter an imagined ignorance and apathy in the broader population by encouraging a broad-scale educational awakening. As a 2006 magazine cover from *The Nation* urged, Americans need to "wake up" and "pay attention" to what they eat.

Both within and beyond SGCP projects, such a fact-based framing furthers the common belief among many alternative food activists (including, as I will show, many of my interviewees) that a lack of individual knowledge is *the* major obstacle in pushing the movement forward, rather than the structural inequities or social hierarchies that prevent equal food access and attachment. "If they only knew where their food came from . . ." the lament goes, as though knowledge itself would solve the problems of our food system (Guthman 2008b). This framing perpetuates the idea that some of us are "in the know" about what is best to eat, while "others" (often framed as minorities and/or the poor) are either ignorant dupes, or worse, careless consumers whose individual choices further the destruction of the planet and local communities. Moreover, when we consider that educational status, racial identity, class standing, and cultural attachments all determine how we come to *know* food (differently), this rigid, fact-based push for one particular type of "correct" knowledge also perpetuates a number of broader hierarchical structures—racism, classism, and sexism, for example.

Significantly, in tacking human health (and no doubt, "correct" body size) onto the list of food-based concerns, the push to become "aware" of our bad food choices has now gained even more momentum. Through mechanisms of nutrition education, such as food pyramids or good food/bad food lists, distinct "shoulds" of eating are now encountered not only as political or ecological necessities but also as scientific and biological truths that need to be learned and understood by everyone, everywhere (but particularly those who are deemed to be lazy or asleep). The scientific backing of alternative foods therefore helps to universalize and naturalize the tenets of alternative consumption, pushing local eating as a natural and apolitical act, rather than something that is social and questionable (a problem that I discuss in more depth in Chapter 1). In some ways, this push has been a very effective strategy in promoting "healthy alternatives," but it has also precluded attention to diversity and difference (or lack thereof) within the alternative food movement.

As my interviews with alternative food activists suggest, in light of the push towards education and awareness, many activists have come to consider schools to be the ideal place to take on this labor of "enlightening." As one activist put it, in schools you have a ready, "captive audience" of students who are relatively easy to convince. But, the implication here is that the learning that goes on in schools is a unidirectional, "banking" system type of education where information flows from holder to recipient (hooks 1994). That is, there is an assumption, held by many, that certain people (experts/teachers) hold the correct information about how best to eat, while others (novices/students) need to and will become educated. Alongside this assumption is the implication that the students themselves are empty vessels, with little or no agency, and with no significant cultural attachments or social positions to "bring to the table."

This assumption carries over from the broader alternative food movement, where alternative food leaders and nutrition "experts" are lauded for holding correct and universally relevant food knowledge, while the rest of us—and specifically the economically and socially disenfranchised, as well as the larger-bodied—are, by contrast, considered naively in the dark. Beyond perpetuating the myth that food choice is ultimately a matter of individual preference and behavior, as several food scholars have shown, the result of this rigid, universalizing approach to alternative/ healthy food has been a "chilling" of those less racially, culturally and economically connected to the movement towards its overall agendas and goals (Guthman 2008b, Hayes-Conroy and Hayes-Conroy 2010a). In other words, the insistence of alternative food activists on one right way to eat has led to the reproduction of a racially, culturally, and economically homogeneous social movement that in many ways continues to be ineffective at reaching out beyond (let alone addressing/ dismantling) such lines of social difference. For example, we can consider that many of the spaces of alternative food—farm markets, community gardens, whole foods stores, and now (as I will argue) SGCPs—are often socially coded as middle-class, white (Slocum 2007), and also healthy, "slim" spaces. The ramifications of this sort of coding for the effectiveness and motivational reach of SGCPs are a large part of what I set out to understand in this research.

Questioning the difference of sensory education

Despite the strong social connections between healthy, alternative eating and middle-class, white, slim culture, it is important to note that many of the alternative food leaders involved with SGCPs insist that food is our "common ground," and that eating is a "universal experience" (Waters Online). Such assertions were in effect my initial impetus for taking on this research, since I viewed these statements as fundamentally detrimental to the success and progressiveness of the alternative food movement as a whole. Beyond the mundane fact that we all have to eat, I came to discover that the reasons for this misguided understanding lie in the ways in which sensory education is imagined to work. SGCPs are often considered to be equally accessible to all students, or even more, to be a great equalizer among students from different cultural, racial, or economic backgrounds. The imagined key to this equality is, quite interestingly, sensory perception. As one teacher from the Berkeley SGCP explained to me: "The value of sensory based [education] is that it is accessible. You are not doing lectures, you are giving equal access and equal opportunity for students to engage with food" (Teacher, CA). In this way, SGCPs are conceived as sensory education initiatives that will magically unlock "correct" food behavior by compelling students to use their senses. Some leaders imagine SGCPs to be equality-producing because they *only* or at least primarily require students to taste, smell, and touch food, as opposed to intellectually or rationally engaging with food, which may be more difficult for some students. But, there are at least two problematic inferences here: first, that prior to entry in these programs (some) students (i.e. the ones going to McDonalds for their food) are not using their senses; and two, that students' sensory perception exists as a natural/essential category that is both prior to and distinguishable from their social experiences and intellectual development. It is because of such assertions and assumptions that my interest in SGCPs first took on a "visceral" feel. I was interested in understanding how SGCPs motivate children to make changes in their eating habits, if they indeed do so, and also how this motivation (or de-motivation) differs across lines of social difference. Ultimately, the question of motivation was to me a visceral question—involving the material sensations and experiences of the human body. But it was a question that was also in no way apart from the realm of social (re)production.

School gardens and other sensory food-based initiatives have been shown to be successful in motivating many kids to change their food habits (Morris and Zidenberg-Cherr 2002), but there are a lot of questions left unanswered. In what ways are these programs motivating to students? How is motivation accessed and experienced by students differently? What types of attachments or ideas motivate students across different means, and towards (perhaps) different ends? How far or how long does this motivation carry, and what obstacles exist to the continuation of motivational experiences in students' future lives? Most importantly, what does this motivation *do* (if anything) to address and dismantle the broader social hierarchies of race, class, gender, and age (among others)?

Furthermore, although motivation is often considered a positive visceral experience, it is important to recognize that this is not always or necessarily the case. In order to expand beyond the current (homogeneous) bounds of the alternative food movement, food activists need to begin to consider how and why alternative food can be "animating" to some and "chilling" to others (Hayes-Conroy and Hayes-Conroy 2010a). Moreover, considering the social pressures on both men and women to maintain an acceptable body size, we must remember that motivation itself (even as animating) is not always or necessarily a politically progressive process. Students (and their parents) can be motivated in ways that perpetuate, rather than disrupt, social hierarchies and material inequities. If we are left to assume that SGCPs "magically" unlock correct food behavior (as is often repeated by some of the leaders of such programs), then we will fail to account for the ways in which bodily attachments and social experiences are central to the circulation of power, and to the (uneven) distribution of empowerment, within the realm of alternative food.

Repetition, disruption and agency

The irony of the oft-repeated "magic" of SGCPs is that, as numerous teachers and volunteers can attest, the space of sensory education takes careful attention and purposeful preparation—sensory education does not just happen organically. Whether planning the garden rotations, the recipe lists, or the table arrangements, SGCPs take place only through a great amount of determined effort—a fact that I both witnessed and was a part of daily during my fieldwork. Once all of these conditions are in place, however, as some SGCP leaders claim, the cupid-esque "magic" is then said to take over: "The only way they will change their [eating] habits is to fall in love . . . I want [the kids] to fall in love. You can prepare the romantic circumstances [for this to happen]. And you can get some really good matchmakers" (Interview with SGCP leader, CA).

In encouraging kids to "fall in love," one of the most frequently repeated pedagogical strategies of the SGCP teachers was, quite simply, repetition. As one teacher explained to me, "You introduce a food again and again, you normalize it, and then it does not appear so foreign." If the end goal of SGCPs is a return to natural eating via the senses/magic, then such claims and strategies do not seem very problematic. However, if we consider the sheer diversity of cultural differences and social experiences that children inevitably bring into a food education classroom, these statements could be read more along the lines of assimilation or cultural imperialism. Further, we might ask, what happens when students magically fall in love with certain foods, only to find that these foods are inaccessible to them culturally or economically outside the confines of the school walls?

These were the types of questions and problems that I set out to tackle as I began my SGCP research. Clearly, I was unsatisfied with the narrative of a magical return to nature. Like other scholars who are critical of the power of nature discourse (Latour 1993, Cronon 1995), I came to view these explanations as eschewing important political and social tensions that ultimately keep the alternative food movement from expanding in progressive ways (meaning in ways that disrupt social hierarchy). My background in feminist theory led me to question such nature narratives particularly in terms of social difference, leading me to search for the ways that gender, race, class, and age-based hierarchies influence students' experiences and trajectories of eating.

Yet, as I conducted my research, I was struck with the recognition that this critique was also not enough—that it was not the whole story. There is a certain lack of agency in both the nature narratives and the socially attuned re-framing, a certain amount of prescripted teleology that did not allow room for the real complexities and ambiguities of food, eating, and life that I witnessed within the SGCP classrooms. Whether through nature or social construction, these stories did not adequately explain how students and teachers and parents, in interacting with each other in bodily ways through food, have a role in affecting outcomes that are far from certain. What's more, neither of these stories recognize nor sympathize with the complexities and contradictions that many progressive activists face on a daily basis (for example, daily ups and downs, grey areas, tricky decisions, partial solutions, long hours, financial worries, etc.). To be sure, in food activism there is always the tension of working both within and against the "system" (Raynolds 2002). Thus, it was important for me to enter into the educational spaces of the SGCPs with an eye towards both repetition and resistance, duplication and disruption, injustice and agency.

In particular, I came to view the SGCP classrooms as spaces in which the living out of food-based identities is occurring continually, haphazardly, and also purposefully. This is not to say that this does not happen in other educational contexts or arenas; but SGCP are by design, hands-on educational projects, and thus are good places to explore the kinds of questions I wanted to ask. With these ideas in mind, my research became a project of becoming attuned to the realities and inequities of social difference within alternative food, while at the same time allowing for the inevitable possibility that "life becomes resistant to power when power takes life as its object" (Deleuze and Guattari 1987). That is, SGCPs are indeed structured spaces in which the current inequities of power within the food system can be experienced and reproduced, but they are also lived spaces in which embodied life experiences can contradict the norms and groupings of social life, allowing for different outcomes to emerge. In recognizing these dual certainties, this book is written with the knowledge that SGCPs are both complicit and resistant to power, and in the hope that they can become increasingly a project of the latter.

Organization of the book

The rest of this book is organized into three parts, each comprised of thematic chapters that explore different but interconnected issues that I encountered in my SGCP research. In Part 1, the reader will find three chapters that focus on visceral ways of coming to know food. Chapter 1 discusses the concept of the visceral in more depth, detailing various theoretical contributions to thinking about bodily transformation, food attachments, affective experience and the material body. Specifically, I use this chapter to describe why the concept of the visceral, drawing from these various theoretical contributions, is useful for understanding and analyzing the impacts of SGCPs. Chapter 2 then addresses the question of what it means to do visceral research. This chapter discusses more details about my specific SGCP research endeavor, including what it looked like (and felt like) to do research on and with the body, and also what this meant for the kinds of results that I, as a bodied researcher, sought to produce. Chapter 3 takes up the question of what it means to know food, and again explores this question through the lens of visceral experience within SGCPs. More specifically, in this chapter, I highlight how a focus on an enlightenment model of food education can be damaging and threatening to the alternative food movement at large, and to the case study SGCPs particularly, by legitimizing certain, socially coded ways of eating over others. In addition, I also explore how the production of food knowledge in SGCPs and beyond can encourage more creative and constructive ways of relating to food that contradict or reverse the effects of this enlightenment model.

Part 2 moves on to highlight four different, *yet always intersecting* axes of social difference—class, race, gender, and age. The chapters in Part 2 draw extensively from the empirical data that I collected during interviews, participant observation, and other ethnographic methods. Chapter 4 brings the questions of knowledge production and viscerality from Part 1 into a study of social class difference, examining the ways in which the case study SGCPs both reproduced and resisted the social hierarchies that are typically associated with socio-economic status. Chapter 5 follows the same overall arch as Chapter 4, this time highlighting racial difference as a salient analytical lens. Gender differences constitute the lens of Chapter 6, focusing again on the ways in which the SGCPs offered opportunities for both reproduction and resistance of normative social categories and behaviors. Chapter 7 then examines the idea of "kid food" and age-based differences in relation to food identities, exploring the ways in which these issues were taken up by and reproduced through the day-to-day operations of SGCPs.

Pulling together the class, race, gender, and age-based analyses of Chapters 4 through 7, Part 3 highlights the implications of this work for food policy and practice, focusing especially on both activism and education. Chapter 8 discusses the pedagogical work of the SGCPs, considering what it means to be attentive to visceral difference in food-based education. This chapter examines the ways in which we can see SGCPs as embodied spaces of learning, and details the mechanisms through which SGCPs may be pushed to become more progressive and less hierarchical spaces of visceral education. In the concluding chapter that follows, I briefly summarize the results and implications of my SGCP research overall, and offer some possible avenues for continued food-based research and activism.

Part 1

Table settings

1   Exploring visceral (re)actions

We are our molecules; our deepest fears, joys, and desires are embodied in the chemical signals of our neurotransmitters. But we are also creators of meaning, making up—and made out of—our histories, our idiosyncrasies, our crazy plot-lines, our unpredictable outcomes. How are we to make sense of the fact that we are both?

(Brison 2003)

As social science research reaches further into the human body as both a theoretical scale and an empirical location, philosopher Susan Brison's question seems to me at the crux of such inquiry. Her question is not necessarily new to fields like geography, women's studies or philosophy, but it importantly places the body at the center of our social and spatial questions—including questions about the food we eat. How are we to make sense of the fact that we are, like the land-scapes that feed us, both biological and social, molecules and meaning, matter and discourse?

I want to begin to answer this question first by looking at the concept of "nature'" and its relationship to the human body. Nature as both an idea and a material thing is given particular form through the human body, and especially through bodily experience. Notions of naturalness and pre-social purity abound in discussions of bodily sensation and judgment, and these sensations and judgments are in turn lent a sort of felt legitimacy through their understood status as "natural." This interplay between (often subconscious) cognition and physical bodily sensation is at the heart of what I conceive of as "the visceral"—the fuzzy place in the body where molecules and meanings, or matter and discourse, collide. Just like nature itself, the visceral is far from a pre-social phenomenon, and yet it is often discussed and experienced as such. In fact, many people use the terms natural and visceral interchangeably when discussing bodily sensations and judgments. Visceral, it seems, has emerged in the English language as nature qua bodily feeling. But despite the limits and dangers of this current usage, it is never-theless a useful concept, worth both critiquing and retrieving as an analytical tool.

The visceral as (not so) natural

In popular culture, visceral reactions are commonly imagined (and experienced) as "natural" in so much as they are assumed to derive from biological and chemical processes within the body. Importantly, these processes are frequently deemed to be apart (somehow) from the body's role as author, thinker, and social actor, eschewing the role of cultural experiences and social differences in the formation of visceral reactions. In relation to SGCPs, a good example of this is taste. Taste is a visceral reaction with many dimensions, and can refer both to broad cultural and aesthetic partialities, and also to preferences for specific cuisines and culinary specialties. In both the broad and specific meanings of the term, taste has been shown to be influenced greatly by one's social positioning (Bourdieu 1984). Yet in common language, many of the SGCP leaders I talked to discussed taste (for food in particular, but also for food practices like eating slowly or eating at a table) as ultimately a pre-social matter:

> The actual taste of the ingredient is not the determining factor, because you know what? A McDonald hamburger tastes like crap. It is not taste, it is marketing, it is social pressure. *Once you get over all the rest of it*, a fresh picked tomato actually tastes better.
>
> (Leader, NS, emphasis added)

> *Activist*: Cheetos are never "good" food. They are a product of production. Who you are does alter your consciousness, but . . .
> *JHC*: Not your taste buds?
> *Activist*: Hmmm . . . I don't know.
>
> (Activist, CA)

In these examples, individual preferences for food are imagined to be complex and are acknowledged to be muddied by social pressures and capitalist reproduction. But, "once you get over the rest of it," our material bodily impulses to choose certain foods over others, as expressed through the biological work of taste buds, for example, are assumed to exist as a pure, pre-social molecular force towards (the right) behavior. Importantly, because these SGCP leaders consider taste in such pre-social terms, they also regard eating as, ultimately, a universal experience. In so far as SGCPs encourage sensory-based learning (as opposed to intellectual learning), the programs are imagined and promoted as spaces that give "equal access and equal opportunity for all students to engage with food," particularly because "all you have to do is use your senses" (Teacher, CA). The implication here is that through sensory perception alone, all students will naturally come to the same (visceral) conclusions about what tastes good.

> You are being educated by osmosis, rather than having to sit down and evaluate, it is happening in a different way, through the senses.
>
> (Leader, NS)

It is a beautiful thing to see that transition [to alternative food] happen, and it happens naturally.

(Leader, CA)

In the above quotations, nature is imagined to work through the body to bring about a certain set of fixed visceral judgments, which are interpreted (in being "natural") as ultimately good, right or beautiful. In contrast, tastes for foods *other* than alternative are therefore implied to be bad and against nature, which advances the educational and political goal of SGCPs to help students "get over the rest of it." The implication is that, at its pre-social core, any*body*'s visceral reaction to alternative food will be the same.

Examining such claims is important to understanding how these SGCP leaders imagine the programs to work through the visceral body, and also how they interpret SGCP outcomes or successes. Particularly, it is important to recognize that these claims are used to frame the visceral goals of the alternative food movement as essentially apolitical, and thus inherently unproblematic. In other words, the healthy eating objectives of SGCPs appear ultimately as a common good for all students, in so far as they are presumed to be nature's intended outcome. They also quite interestingly appear as the indisputably tast*iest* choices, to the extent that nature is imagined to really determine "actual" taste. Thus, in Berkeley, for example, the oft-repeated promotional sound bite for SGCPs tells us that the food kids eat at school is simply "delicious," (see Alice Waters' "Delicious Revolution;" Waters Online), begging the question, why don't the students get to make this call? Rather than allowing the students themselves to assess the deliciousness of the food, this claim is made a priori, with nature's unquestionable goodness/tastiness standing in for empirical evidence from those who actually eat/taste the food. In contrast to this, however, stand the many student complaints that I heard while working in the SGCP classrooms about the changes in school food, particularly at the level of the cafeteria:

Last year they changed our lunch program, so that it is all organic. But I stopped buying it because it doesn't taste good. I mean, I want organic, but I want it to taste good too. Like cardboard and pears? Not cool . . .

(Student, CA)

We like pepperoni pizza, but we can't have it because it is not healthy. [The teacher] told us that it was the school board that made that decision, but I [don't understand] because [other schools] still get to have it.

(Student, NS)

Beyond disallowing the possibility that students (need to) have personal agency in the development of taste preference, claims about the "natural" and "actual" correctness of SGCP food also serve to rationalize the superiority of alternative food preference over other, nonconforming tastes (e.g. a taste for pepperoni pizza). This rationalization is particularly problematic given that the alternative food movement is *also* frequently characterized (even by many alternative food activists themselves) as economically elitist, racially white, and culturally European. In this sense, any assertion of a pre-social quality to SGCP food is not only questionable as a supposed fact but ultimately condemnable as a potentially pernicious rationalization of the imagined supremacy of rich, white, and western forms of knowledge and valuation over other forms of knowing and assessing food. These dilemmas clearly illustrate that the boundary between the nature and culture of taste in SGCPs is far from clear cut, and thus that the goals of alternative food are by no means beyond the bounds of social critique. Indeed, we might consider that what seems like untainted nature to some can feel more like cultural imperialism to others.

Visceral difference and visceral change

Acknowledging that visceral judgments like taste are far from pre-social, how are we to understand the (potentially positive) power of the visceral within and beyond alternative food? It seems that the metaphorical appeal and indeed the material power of the visceral in many cases come from its imagined (and improbable) attachment to "pure" nature. Certainly it is because the visceral often *feels* so natural—so innate—that many activists and eaters will trust it as a source of great reason. After all, when in doubt, we should just listen to our guts, right? But, if our guts are nature's unhindered pulpit then why don't we *all* crave to eat from a farmers market, or an organic garden?

My point here is that visceral judgments *do* influence the trajectory and outcome of many social and political phenomena—including what we eat at our tables. This is a claim that I wish to expand upon extensively in this and subsequent chapters. But, as the sheer diversity of our tastes and preferences would suggest, these visceral judgments are certainly not "pure" nature at its essence. Instead, then, I want to suggest that we begin to imagine "the visceral" as a collection of learned bodily habits towards certain bodily feelings, desires, and actions. These habits are not static and universal but are materially produced and reconfigured through various biological and chemical flows that are situated within people's different social, cultural, and political contexts. They are our daily living out of the messy interaction, the ongoing chaotic relationship, between Brison's (above) meaning and molecule. In this sense, visceral reactions defy the oft-presumed boundary between nature-as-molecule and culture-as-meaning, suggesting that it is instead the *intersection* of these supposed pure forms that propels both social and natural reproduction (Latour 1993). In this hybrid understanding, inclinations towards visceral reactions inevitably differ from person to person, group to group, and context to context—and thus, instead of "pure" universal response (based on unitary, pre-social forces), we can encounter contextualized instances and patterns of often inequitable and conflicting visceral difference, in which physical bodily matter is always already socially and differentially articulated (Latour 2004). This, of course, does not make visceral judgments any less real, any less tangible, or any less natural. But it does make them more open to critique, and also more viable as instruments of social analysis and social change.

Given such complexities, to talk about viscerality is (admittedly) to try to specify the unspecifiable—a *something* that is constantly shifting, deeply contextualized, and haphazardly produced. It is truly difficult, if not impossible, to pin down exactly why we taste what we taste, or feel what we feel. These bodily judgments are at once internal and external, historical and immediate, probable and unpredictable. The larger spatial and temporal arrangements that at a moment produce within a body a particular craving, for example, are too numerous and disorganized to identify; and the related internal mechanisms of perception and judgment that attempt to filter and organize these arrangements are similarly copious and complex. For these reasons, as I discuss visceral reactions in this book, I will not attempt to draw direct causal linkages between certain social circumstances and specific food habits, nor will I attempt to argue that cultural differences can assuredly forecast taste preference. These stories may exist within my case studies, but if so, they are only ever partial descriptions. I want to encourage an understanding of visceral reactions in SGCPs that is much more fluid than this.

Still, I also want to argue that visceral reactions, despite connotations of animal instinct and uncontrollable response, are also not entirely out of our reach—nor the reach of others. The visceral realm is not thoroughly mysterious and unknowable, even in its complexity and specificity. Certainly the success of the advertising industry in eliciting product-based visceral responses is testament to this fact, such as when children declare that carrots packaged in a McDonald's wrapper taste better (Robinson et al. 2007). Indeed, as geographer Nigel Thrift warns, knowledge about the creation and mobilization of bodily affect is being deployed knowingly and politically, and "mainly . . . by the rich and powerful" (2004, 58). Nevertheless, progressive activists also can and do pay attention to the ways in which bodies come to sense and judge and (re)act (Hayes-Conroy 2009, Hayes-Conroy and Martin 2010, Clough 2012). Visceral judgments are not passive, unthinking, "reactive" responses but instead learned and embodied behaviors that involve a series of active and often purposeful (though not always conscious) decisions. While not fully predictable, therefore, visceral reactions are certainly manipulable, and thus open to many possibilities of *re*-activation. I therefore want to approach viscerality as something that we can also learn *to use* in effecting personal and social change at a variety of scales. This is not to deny that other avenues of social change are important (e.g. policy or institutional reform), but rather to recognize the visceral as present both beside and within other mechanisms of change. To think of visceral reactions as embodied socio-spatial productions—material relationships through which we develop diverse inclinations towards certain behaviors, actions or feelings—is to take a step towards understanding visceral response as a mechanism of daily political struggle.

In the sections of this chapter that follow, I explore in more depth how academic scholarship has contributed to what I refer to in this book as "the visceral." The examination is not exhaustive, and is also not meant to underscore minute differences. Scholars have taken interest in the body for a variety of reasons (e.g. cultural analysis, health studies, political strategy), and have discussed the body using a variety of terms and concepts (e.g. embodiment, performance, affect, and also "the visceral"). The goal of this chapter is less to explain how these interests and concepts differ as to acknowledge the progression of important ideas in regard to bodies, to explore significant areas of convergence within body-centered scholarship, and to flag places where the body has been discussed in ways that I find particularly useful for examining SGCPs. Recognizing that what any of us are trying to describe when writing about bodies is constantly in flux, I believe that it is important to recognize that there is no one right way to think about bodies, though there certainly might be mistaken or obstructive ways.

Turning (back) towards the body

Historically, academics have shied away from bodies and visceralities as topics of scholarly interest. There is a well-worn history of this in the academy, critiqued roundly and rightfully by corporeal feminist scholars, among others (Grosz 1994, Birke 2000). Indeed, bodies and their sensations and judgments have long been associated with women and notions of femininity (e.g. being ruled by emotions), while men and masculinity have been linked to rational, cognitive thought, and thus to legitimate intellectual inquiry (Brison 2003). In addition to such sexist reasons, however, academics have also turned away from the material body for much more understandable, and even productive reasons. For example, within geography, early theories of "biological determinism" had sought to explain cultural differences in individual behavior through differences in genetics and racial categorization. As these racist and sexist theories became rightfully discredited (Livingstone 1993), social theorists, including many feminist scholars, were hesitant to return to any body-centered explanations, focusing instead on analyses of social disenfranchisement and economic inequity.

In spite of this "turn away" from the body, however, in recent decades there has been much important scholarship on bodies and embodiment, including especially from corporeal feminist scholars. Much of this work has drawn attention to the body as it constructs and is constructed by our social, economic, and political worlds. Feminist scholars have pushed to recognize the body not as an isolated biological object, but as a location for the practice of social values, beliefs, and norms. Many of these scholars remained consciously distant from theories of the body as biologically prescribed, and worked instead to forcefully deconstruct any biological essentialism within social theories of embodiment. Despite the importance of such theorizing, though, other feminist scholars have also critiqued social constructionist work on the body, arguing that this work has not given enough attention to the *materiality* of the body—to the actual "messiness" and "fluidity" of bodily practices and processes (Longhurst 2001). This critique is important in the attention it brings to the material body as it develops and is developed within various social contexts. Indeed, in many ways, social constructionism has tended to "render the body incorporeal, fleshless, fluid-less, little more than linguistic territory" (Longhurst 2001, 23). However, while social constructionist arguments have not necessarily drawn attention to *how* the body is materially in flux, I would argue that much of this work has inspired us to rethink our assumptions about a fixed biological self.

For example, feminist scholar Judith Butler's (1990) *Gender Trouble* was instrumental in this type of anti-essentialist work, and is particularly notable because it destabilized feminist theory to its core. Butler took biological essentialism in women's studies head on, deconstructing the very subject upon which feminism is (thought to be) founded—that is, the "woman." She wrote, "The feminist subject turns out to be discursively constructed by the very political system that is supposed to facilitate its emancipation" (1990, 4). In other words, Butler suggested that not only is gender socially constructed, but the very category of female itself is also discursively produced. The sex/gender distinction is therefore no longer valid to Butler—sex is not a "natural" biological category, determined and fixed by our essential, pre-social being. In fact, she maintains, there is no such thing as a pre-social body—bodies are always, already discursive (Butler 1993).

In this work, the resistance to biological essentialism is powerful. Butler takes important strides towards rethinking the body and bodily experience as always, already social. Moreover, she eschews biology for an important reason: she wants to argue that the discourse of (fixed) biology is a powerful force in controlling and compartmentalizing bodies—in precluding what certain bodies can be and become. While we might be left to wonder how the *material* of our bodies (in nonfixed terms) can act back upon this social discourse, or how our bodies develop certain visceral assessments of categories like male and female, right or wrong, and good or bad, Butler's work undoubtedly tells us much about the broader social relations out of which our bodies develop certain judgments and habits of feeling.

What is perhaps most important about Butler's work is the possibilities that it opens up for rethinking the role of the visceral in social practice. Butler moves from explaining that bodies are controlled and normalized, and that social experiences and bodily performances are perceived as "natural," to offering insights into how we might begin to become aware of these processes of normalization/ naturalization, and thus begin to control them ourselves. Her strategy of attack is also largely discursive—but the imagined result is certainly not. Butler explains, "[w]oman itself is a term in process, a becoming, a constructing that cannot rightfully be said to originate or to end. As an ongoing discursive practice, it is open to intervention and resignification" (1990, 43). The attention to process is particularly notable here—bodies are not just inscriptions of culture, but continual works in progress, of which we can take an active part.

The language of "performance" in Butler's work has also been especially important to geographic and feminist scholarship on bodies and embodiment. For example, feminist geographer Catherine Nash offers the metaphor of performance to explain that "women and men learn to perform sedimented forms of gendered social practices that become so routinized as to appear 'natural'" (2000, 654). This metaphor allows us to imagine both that we reproduce social norms through specific and repeated bodily acts (e.g. performing femininity through eating certain foods, or engaging in certain tasks in the kitchen), and that, as we learn how to identify these mechanisms of reproduction, we can begin to resist and change such roles through our bodies as well (e.g. by opening up to other types of performances, or other kinds of food attachments). Thus, as Nash goes on to suggest, the metaphor of choreography is "perhaps better than performance in conveying the idea of codes, traditions, and conventions that people reproduce, rework, parody, or upset—the constraints and rules as well as the possibilities for subversion." In other words, *because* the body is social (read: changing), it is also (potentially) retrievable as an instrument of social change.

Feminist geographer Linda McDowell (1995) similarly noted that in the service industry, the performance of gender roles is a requirement for employment. Though professional jobs are often seen as "mind" work, McDowell tells us that jobs are increasingly characterized by what she recognizes as "body" work: by sexualized performance, careful disciplining, and self-surveillance. For McDowell, this recognition is not an end in itself but is the beginning of a broader political project to understand and take control of bodies in social reproduction:

> The body is the primary site of social experience. It is where social experience is turned into lived experience. To understand the body we have to know who controls it as it moves through the spaces and times of our daily routines, who shapes its sensuous experiences, its sexualities, its pleasures in eating and exercise, who controls its performance at work, its behavior at home and at school and also influences how it is dressed and made to appear in its function of presenting us to others. The body is the core of our social experience.
>
> (Fiske quoted in McDowell 1995, 76)

While neither McDowell nor Nash explicitly imagine the role of the internal, material body in these accounts (e.g. through muscle memory, habit formation, or physical movement), certainly they want us to understand the body as a material site through which we *physically* (*sensuously*) experience social interaction. The involvement of the material body is at times more passive than active in these accounts (with "pure" social forces as the active agent), but at other times we can begin to imagine how the material body can be harnessed (even disciplined) in countering oppressive agents. We can imagine, for example, how things like pleasures or tastes might be reclaimed or "taken back" by a sort of visceral playing and retraining of such bodily habits (Hayes-Conroy and Hayes-Conroy 2008). There is an important implication in this work, that as we begin to pay attention to how our material bodies are socialized, and how this socialization influences our lived experiences (such as through eating), we will be better positioned as students, teachers, and activists to "choreograph" resistance to oppressive social forces (Nash 2000).

Resistant bodies and material change

In regard to SGCPs, such practice of political resistance through the material body requires an examination of (the changing nature of) taste and food habits. But, how are we to specify the role of the material body in producing visceral judgments like taste preference without conceding to biology as a static and essentializing science? We need to begin to describe with more specificity how the material body is able to change, and also how material conditions and relations influence the way that change happens. To do this, I return to Brison's question that opens this chapter. For Brison, the embodied hybridity of molecules and meaning is not only an intellectual and political inquiry, but also a journey that is essential to physical bodily *healing*—a specific and purposeful type of bodily transformation. In her book, Brison recounts her personal recovery from the trauma of rape as a material process of biological change, one that involves simultaneous intellectual and physical struggle in the aftermath of violence. She describes how her flashbacks, goose bumps, rapid heartbeats, sensations of panic, and other bodily conditions slowly begin to shift and dissipate as she actively labors through the cognitive (meaning-making) and corporeal (muscle-building) journey of self reinvention. In other words, her body changes (heals) through a negotiated yet disorderly process of "conscious" visceral *re*-acting in relation to both the social definitions and the physical training that reproduce her post-trauma self.

Brison's work allows us to recognize at least three important characteristics of the material body that are essential to research on SGCPs. First, the body is ever-changing, a "process-of-being," and thus always in material flux. Bodies (re)act (differently) to social and environmental stimuli in ways that constantly shift the sensations we experience and the judgments we make. In Brison's account, it would be hard, if not impossible, to draw a definitive line between mental/cognitive and physical/bodily healing; the visceral shifts that she labors through happen in the spaces in between. Similarly, in a cooking or gardening classroom, the boundary between sensory education and cognitive engagement is not clear cut; students' visceral (re)actions to food take place as a hybrid of the two forms of development, learning what is "good" to eat. In this sense, the material flux of the body through taste and habit formation is not separate from the body as a social construction. Rather, social norms and identities are given material legitimacy, and also resisted, as they are (re)evaluated through conscious and subconscious bodily judgments and reactions. It is important to recognize that the material body is also fluid in this process, constantly shifting in terms of how, when, and to what it (re)acts— or to what it senses/tastes as good. Whether in regard to the production of tastes or the process of healing, performances of social norms and roles change alongside physical changes in bodily reaction and judgment precisely because the two are not separate processes.

Second, despite the importance of this recognized fluidity, bodily change is not (often) drastic but rather incremental or developmental. Although fantastic or traumatic events can radically and rapidly alter one's visceral experiences, in everyday life activities like eating, or in the aftermath of such weighty events, change most often comes slowly. Visceral reactions therefore *feel* natural (read: fixed) for good reason: bodies form habits of feeling and perceiving that become subconscious and often increasingly viscous (Alcoff 2006). Our response mechanisms become accustomed to firing in certain ways, and it can become harder and harder to "feel out" new ways of judging, (re)acting, or tasting. Indeed, habits lend a sort of comfort and stability that is not always easily or contentedly altered, neither in SGCP classrooms, nor in other social spaces. In addition, while there is also much haphazardness and uncontrollability within our larger *external* world, there is furthermore a lot of undeniable "stickiness" (Slocum 2007, 2008) in this world as well—in terms of, for example, the entrenched-ness of hierarchical social norms, definitions, and expectations in regard to "correct" eating, or in terms of the continued inequities and inefficiencies of our health care and food systems at large. Thus, although life is always changing, we also must be careful to not assume fluidity as an inevitability. In Brison's account, she labored through high levels of internal and external "viscosity" (Saldanha 2005). And, she approached fluidity not so much as an inevitability, but as a developmental and political practice.

Third, Brison's work reminds us that (despite high viscosities) we have at least some ability to consciously affect and direct such change, through deliberate engagement with the socio-spatial and material contexts that influence a body's re-creation. Even though we cannot fully dictate (or even fully know) why our body feels or (re)acts in certain ways, we can consciously have an influence on these processes. For a food-related example, we can and do develop tastes for certain foods, to the point where previously bad tasting foods can begin to taste good (i.e. we can develop a "taste" for them). It is hard to determine the exact causal pathways of such a development, and yet certainly many of us can and do frequently work towards new tastes and preferences (for personal, practical or political reasons). Still, as Brison's own experiences caution, consciously recognizing and reacting to our bodily tendencies or viscosities takes more than simply intellectual will, and more than just a flexible living out of life. It takes conscious individual and collective effort to understand our visceral reactions, it takes deliberate collaborative work to change unhelpful or inequitable socio-spatial conditions that preclude progressive change, and above all, it takes time to do both of these things (a claim with which many alternative food activists agree). Brison wrote her book 10 years after her traumatic experience—and her process of healing was (perhaps is) still ongoing.

Relational agency and political action

Beyond Brison, many other feminist theorists have articulated similar(ly political) projects of bodily transformation (among them: Johnston 1996, McWhorter 1999, Birke 2000). Feminist disability scholars, for example, have been particularly helpful in (re)examining the role of the physical body in the process of progressive social change (Hall 2011, Kafer 2013). And other health scholars, such as Annemarie Mol, have engaged in exciting and path-breaking work that analyzes the various socio-material networks through which bodily health is materialized (Mol 2002), and also made meaningful (Metzl and Kirkland 2010). Queer theory scholars, too, have picked up the question of bodily change, looking at processes of animation, for example (Chen 2012), or how bodies become physically oriented (Ahmed 2006). What all of these works collectively point to is a desire to better understand the role of the material body in the various processes of political action and social change.

In *Bodies and Pleasures*, philosopher Ladelle McWhorter (1999) describes a political process of bodily change that involves the training of muscles in dance as part of a conscious effort to reclaim ethnic heritage. Although McWhorter does not, as Brison does, frame her work as related directly to healing, the process of individual bodily transformation that she describes certainly links her personal, bodily well-being to broader social and political struggle. If ultimately such inquiries into the contextually situated, minded body stem from a desire to effect healthy and equitable social change, then these examples also urge us to reassess our conception of personal agency. Particularly, in engaging with notions of the body as fluid and changeable, it becomes both practically and politically necessary to put forth a conception of agency that is not individual but relational. It is necessary practically because a relational conception helps to explain ontologically how bodies "become" what they are. And, it is necessary politically because autonomous bodily change, as Brison recounts (2003), is intimately *dependent* on one's broader community and context. Mistaken notions of the body as purely independent can lead to unhelpful calls for individual responsibility and correct choice, such as when food activists or nutritionists call for more "knowledgeable" food decision-making rather than more equitable political and economic contexts.

In the past two decades, such theories of relational agency, including especially Actor Network Theory (ANT), have emerged in the social sciences as a critique of the nature/culture dualism, advancing a notion of agency that is grounded in material interrelations rather than exclusively individual dealings (Haraway 1990, Latour 1993, Whatmore 2002). ANT argues, for example, that events only occur as various actors (or "actants"), both human and nonhuman, come together in physical and discursive connection to produce specific outcomes. This relational or networked theory of agency is often put forth in regional and global explanations of "natural" entities or phenomena (Whatmore 2002), but it is also an important starting place for understanding both the *internal* and *external* mechanisms of the material body—and indeed for recognizing the interconnectedness of these two. ANT allows us to recognize that the seemingly stationary "matter" of the world is in actuality a continual and chaotic set of interactions—a web of relationships that is not fixed, but also not without constraints. In this web, agency is not located in any particular person or thing, but is to be found through the process of interaction itself. Thus, agency (as the capacity to effect change) is not made up of conscious or deliberate acts, per se, but rather emerges out of the messy interaction between material and discursive flows both within and beyond the bounds of the skin. Of course, this is not to say that conscious or deliberate acts are not important, but rather that a person's individual ability to effect change is necessarily mediated through a complex and not fully controllable material–discursive world.

Understanding agency as relational, then, can remind us to recognize the material body as part of the agentic process. We can remember, for example, how neurotransmitters receive and transmit social information (Rose and Abi-Rached

2013), how muscles remember to perform certain acts (McWhorter 1999), how hormones influence emotional experiences (Roberts 2007), how taste buds register sensations of bitterness or sweetness (Hayes-Conroy and Hayes-Conroy 2010a), and how all of these interactions, over time, amount to the development of a knowing, thinking body—to what a body is and what a body is becoming. In other words, we can imagine that tendencies towards certain visceral reactions exist, in the present and future, as behaviors learned in and through the limits and possibilities that our material body offers, as well as the bounds and potentials put forward by our socio-cultural world. In this way, understanding agency within SGCPs necessarily involves the recognition of taste as an ever-emergent process, developed through specific relational contexts and aims, rather than through universal and predetermined "goods."

Food and visceral politics

With a relational rather than individual conception of food-based decision-making, we are led to ask a number of broader questions about the nature of visceral (re)acting, and especially about the relationships between small-scale bodily experience and large-scale socio-political context. In our unequal economic structure, to what extent can food behaviors really be considered choice? What influence does the broader system of "free" trade have on how different people viscerally evaluate food, and also on what food different people can even have the opportunity to evaluate? Where does tradition, or culinary culture, or health knowledge fit in to the production of visceral attachment and desire? How are our habits of feeling or craving reinforced by social norms, gender roles or racial stereotypes? These questions are undoubtedly large, but they are important and central questions to the task of furthering progressive food politics.

From the above theoretical discussions, we can begin to appreciate that individual visceral reactions towards food are neither a fixed biological phenomenon, nor a fully social enterprise. But how can we become more aware of the visceral politics that exist in the hybrid spaces in between? And, how can we give new weight to the personal as political? Feminist scholars have long stressed the political importance of understanding how the micro world of daily, lived experience and embodied practice is articulated in and through our larger macro-political world (hooks 1989, Domosh 1997, Dowler and Sharp 2001, Staeheli 2001). Can a focus on *food* help us to bring this project even further into the body? Many scholars have recently begun taking important steps in this direction, connecting scholarship on material bodies to their research on food habits, cultures, and judgments (Slocum 2008, Longhurst et al. 2009, Hayes-Conroy and Martin 2010, Guthman 2012, Johnston and Longhurst 2012).

Cultural studies scholar Elspeth Probyn's work on food and the body (2000), as well as on emotion (2005), has been especially instrumental in helping to further an academic understanding of visceral politics. In her discussions of food–body interactions, Probyn suggests that eating is a useful place from which to begin to understand the connections between the "micro," molecular world and the "macro" level politics that define and produce social difference. In her view, the key to understanding is eating's sensual, physical, or visceral nature. Probyn explains that while there has been a great deal of abstract work done on the body in terms of the concept of embodiment, "the realm of the alimentary brings these considerations down to earth and extends them" (2000, 3). Through her work, she shows how the everyday, micro scale of daily practices like eating reaches up into the macro-political realm, bringing it to life. She therefore wants to articulate not just a politics *of* the body, but a politics that acts in and through the body, asking: can we envision "a more visceral and powerful corporeal politics?" Can eating help us understand "our visceral reaction to who and what we are becoming?" (2000, 14). Importantly, Probyn wants to focus attention not only on positive responses to food (cravings, desires, attachments), but also negative responses (shame, disgust, fear). She suggests that in paying attention to all of the visceral moments of food, we will be better able to understand not only where we as a society are coming from, but also where we are collectively headed.

Through Probyn's work, we can affirm that different tastes for certain foods are materially developed in the body, brain, and tongue through a particular series of past opportunities, memories, histories, and vacancies. Looking backwards, our visceral responses to food arise out of the same power-laden social networks that Butler and McWhorter, as well as Pierre Bourdieu and many others, have recognized as central to the development of our (always social) bodies. Our present visceral responses can confirm and reinforce these social relationships, but they can also potentially deny or resist them. By studying visceral reactions, we can learn something about the political topographies of our relational world that influence such opportunities, memories, histories and vacancies. But even more importantly, we can learn something about the conditions needed for certain tastes to remain, or to change.

In food studies, then, politics of the visceral is a politics of both society and biology, and is both internal and external to the body. It is necessarily a fuzzy and disorganized politics, where relations are far too numerous to recount or fully explain, let alone effectively control. Undoubtedly, this conception of politics can at first seem daunting for scholars and activists, for where are we even to begin? Yet, it is in this fuzziness and disorganization that many have also already recognized the exciting possibilities for social change. It is the potential and inevitable fluidity of the living human body itself, with its social tendencies, *but not absolute fixities*, that opens us to the potential for newness or change (Hayes-Conroy and Hayes-Conroy 2008). This is part of what gives people the capacity to experiment with the visceral realm, to feel out new ways of being and becoming, and to resist the powers that constrain. And this is also part of what gives SGCPs the opportunity to become political instruments—of the alternative food movement, of nutritional science, or potentially of something far more radical and socially conscious than either of these. The fluidity of life means that play is possible, even somewhat inevitable, if only everyone had the resources, the support, and the space to play.

Bodily affect and emotion

Perhaps the most direct, social scientific engagement in recent years with the internal mechanisms of the changing/changeable body has come through scholarship on affect and emotion (two terms that are sometimes used interchangeably, though have increasingly been deployed as distinct). While not dissimilar to the works discussed above (in fact, there is much convergence), scholarship on affect and emotion has encouraged specific intellectual attention to bodily moments of judgment and movement, and has focused especially on imagining and describing how and why these moments unfold. There are a great many scholars now contributing to this body of work, and far too many to do justice to them all here (Massumi 2002, Sedgwick and Frank 2003, Brennan 2004, Stewart 2007, Ahmed 2010, Berlant 2011). The list also includes many geographers (Nash 2000, McCormack 2003, Bondi 2005, Thien 2005, Thrift 2005, Anderson 2006, Clough 2012), as well as several new affect theory edited collections, which speak to the prominence of this research interest within the social sciences (Gregg and Seigworth 2010, Clough and Halley 2007).

As a whole, scholars of affect have been particularly drawn to network approaches in promoting what some call a relational ontology. Thrift's work on interpersonal affect, for example, puts forward "a different model of what thinking is," recognizing that reflexivity is not just "a property of cognition" but is simultaneously a bodied phenomenon (Thrift 2004, 59). Like Brison, McWhorter and Probyn, Thrift and other scholars of affect want us to recognize that thought (judgment, decision-making) happens within and through bodies (as well as minds). Furthermore, they maintain that bodies affect and are affected by the world in ways that are not easily named or represented. Accordingly, this and much work on affective bodies also falls under the name Non-Representational Theory (NRT) because it involves a different kind of intelligence about the world that is continually unfolding in non-cognitive ways as people move through their daily lives—for example, as people choose what to eat. In this way, NRT seeks to move from epistemological to ontological argumentation.

Scholarship on affect and NRT has emerged in part as a response to more than a decade of research focusing on the politics and culture of representation. Scholars of affect and NRT argue that the continuous focus on discourse, meaning, and social construction in much postmodern social science research has served to gloss over the role of the biological body in the production of thought and action. Much like ANT defines agency as a capacity to effect change that is born out of the relationship between material things, scholars of affect and NRT approach affect not as a matter of one individual body but as a relational event, occurring in and through the interaction of various "bodies" (human, food, music, etc.). This work has relevance to SGCPs in connection to both Brison's concerns about bodily healing and to McWhorter's concerns about political power. For example, non-representational theorists have stressed the need to understand the experiences and knowledge of the body as part of one's therapy (McCormack 2003), and also the body's role in motivating a person to make certain day-to-day decisions (Anderson

2005). Scholars have also suggested that affect is important in understanding and furthering one's political practice (Clough 2012). Thrift, as I mentioned earlier, notes that corporations are now going to great lengths to increase their under-standings of affect, and that this knowledge of affect is being used to further the political ends of the rich and powerful (2004). Thus, he calls for research into the political implications and consequences of this build-up in capacity to manipu-late affect—a capacity that social theorists Michael Hardt and Antonio Negri (2001) and Nikolas Rose (2001) have recognized as the latest version of Foucault's "biopower."

In particular, I find geographer Ben Anderson's work on the practices of judgment and affect (2005, 2006) to be especially helpful in drawing attention to the workings of the material body in regard to decision-making events in everyday life. Anderson outlines a specific theory of bodily affect, using empirical research to describe how "practices of judgment" unfold in and through the material body. In his case, he focuses on judgments about listening to music; yet, Anderson's work certainly has implications for other forms of judgment-making—including (re)actions to food. His discussions of judgment-making within the body help to specify the link between preference/taste and the social world. Anderson claims that sociologists like Bourdieu have tended to discuss this link in a taken for granted way, focusing on the particular chosen *content* of beliefs or tastes, rather than on the material processes of choice itself. While Bourdieu's work has been instru-mental in showing how tastes are not part of the natural and essential self (as many SGCPs leaders allege), Anderson worries that his explanations ultimately serve to trap our understandings of preference and judgment, thereby reifying the same social hierarchies that we want to dismantle. Instead, Anderson wants us to recog-nize that a judgment does not necessarily result in differentiation and hierarchiza-tion, and that it is important that we allow the possibility for something different to occur. Indeed, this possibility for newness is where the political usefulness of visceral materializes. In terms of SGCPs, this would mean that we need to allow space for the performative moment—for the ability of students, teachers, and material life itself to disrupt the social tendencies of taste. In other words, Anderson seeks to re-materialize discussions of life and lived experience in a way that refuses to equate the biological with the essential or the "neuroreductionist" (Anderson 2006, 747).

Making a judgment is therefore not *just* about the social conditioning of an individual, rational human body to be affected in a certain way by a certain kind of music, or a certain type of food. It is a process that is continually varying through an extensive set of material relations that are particular to the time-spaces in which a specific judgment is taking place. The realm of the body that Anderson describes as active in judgment-making is the "more-than-rational," suggesting that there is something chaotic and unpredictable about the practice of judgment-making, and that within this chaos exists the potential for variation and change. Anderson later explains (2006) that affect takes place as "something more, a more to come," because it is accompanied by a real but virtual (potential) "knot of tendencies and latencies," which amount to divergences in what actually becomes (Anderson 2006, 738).

Therefore, in terms of SGCPs, we might think of such "tendencies" as the ways in which visceral reactions towards foods and food behaviors are cemented by social positioning into habits, while, in contrast, "latencies" become the always-present potential for something else to emerge.

Affect as (also not so) natural

Despite the thoroughly social intersubjectiveness of affect, and the potentialities for variation, Anderson's work also confirms what many SGCP leaders have also felt—that judgment-making is typically *experienced* as a purely natural event. This means that as we make judgments in regard to music, or to food, we sense them, instead of relational and contextual, as pre-social, "gut" reactions. Anderson's work seeks to describe why this is the case; he imagines that practices of judgment are a sort of coping mechanism of everyday life, one which allows us to respond to what is immediately confronting us and to achieve a mode of being that enables us to get by, or to continue with other life practices. This coping and continuing is reliant, Anderson explains, on the visceral belief that the judgments we make come naturally; "what underpins the making of a judgment is, in rather circular terms, a visceral, non-cognitive, belief in the veracity of the judgment" (2005, 646). In other words, in order for the world to make sense to us, we must always have a sense of trust in our own tastes and preferences. What we produce in judgment-making is a visceral assumption that our feelings and emotions "express an evaluative stance that is 'natural'" (2005, 653).

This recognition of the *sense* of naturalness or veracity that comes as we viscerally react to different life circumstances is important to an analysis of SGCPs for at least three reasons. First, it lends a compassionate understanding (rather than a blaming antagonism) to the SGCP leaders who assume that "real" taste comes naturally, as opposed to through different socio-spatial circumstances. Second, it offers a more specific interpretation as to why bodily practices of feeling or judging cement into habits or routines that become the likely "tendencies" that scholars like Bourdieu have analyzed. If students experience bodily preferences for eating as natural and essential parts of the self, then it makes sense that the link between social identity and food is solid and hard to disrupt. Indeed, as I witnessed during my interviews with students and parents, food is experienced as an intensely personal subject, any change to which is often felt as a threat or critique of one's character. Third, if we recognize the power of affect to bring about social change, then it is important that we think about how students can experience this sense of trust in bodily judgments even as we actively try to disrupt social tendencies. In other words, if we want SGCPs to encourage changes in eating behavior, how can we create spaces that allow children to make healthy eating "their own"?

As I have already argued, it is vitally important to move beyond the essentializing discourse of fixed biology and pure nature if we are to understand and use affect as a political terrain. Anderson explains that it is only after we recognize the intersubjectivity of affect that we can "open up the social to the neuropolitical domain of hormones and synapses" (Anderson 2006, 747). Ultimately, what I wish to show is that, despite the SGCP leaders' insistence of the pre-social nature of taste, food-related affect *is* political, and is moreover an important political tool. It is political first because it is relational; it is produced out of webs of material and discursive interaction, which are neither equally accessible nor equally controlled by everyone. And it is political second because it is energetic; it involves the enhancing or depleting of the lived space-times through which students discover their agency as the motivation to *do* (or eat, as the case may be).

Thus, to be political with affect, it seems, we must do much more than discuss bodily sensations and judgments produced as "individual" or (even more improbably) "universal" feelings. And we must do more than interrogate affective experience as (not so) purely "natural." Rather, we must examine affect at a number of scales, and through a number of contexts. We must labor against negative affectivities of suffering or depletion that exist both locally and globally, and we must work instead to cultivate the conditions needed for more positive, "empowering" experiences of food and bodies. Most importantly, we must look not only at how affective experiences are labeled or narrated, but also at the *work* that they do in advancing us towards and/or retreating us from progressive achievements.

The language of viscerality

In my mind, one of the great ironies of much of the academic literature on bodies, affects, and relationality is that while it often rests so strikingly close to our experiences and imaginings of the "real world," the texts are often so dense and complex that they take considerable time and effort to appreciate. Indeed, scholars of both affect/NRT and ANT have been criticized for using language that is largely inaccessible to most academics, let alone a lay population. Perhaps to some extent the fuzzy feelings that come from reading these texts are *meant* to mirror the fuzzy feelings of actual interrelational becoming. But, should it not be easier to describe what these authors ultimately imply is our most basic, ontological condition? It is important that, in continuing to promote these ways of thinking about and practicing (with) our bodies, we begin to develop methods of communicating and explaining that resonate with a larger number of people. After all, we don't want to encourage visceral reactions of dismay or detachment among our readership.

Of course, part of the reason for the somewhat convoluted communication is that (the English) language (at least) does not have adequate words to describe the interpersonal agency, socio-material hybridity, or chaotic relationality that such scholars seek to explain (that is, without using these very words). Another reason for confusion is that the movement and fluidity of the social-material processes that these authors seek to explain inevitably make it hard to "name" what is in many ways unnamable. In large part, then, this fuzziness is unavoidable, and even desirable, in that the indefiniteness is both true to present life and open/flexible to the possibility of different futures. Nevertheless, it is worth trying to develop a language through which we can allow this fuzziness to be understood and embraced in somewhat clear, simple terms. In order to do this, I first want to briefly tease out how scholars of affect and NRT use words like feeling, emotion, and affect to signify what are, in their conception, different moments of interrelational life.

Part of the intent of Anderson's (2006) work on affect is to specify a language for talking about affect, feeling, and emotion in order to make it easier to understand what scholars are referring to when using these terms. First, Anderson imagines *affect* as the interpersonal capacity a body has to affect and be affected by other bodies. Affect is the ANT moment of visceral judgment—when agency is found in relationships. Anderson emphasizes that there is not a priori to affect because it is always relational to the past. In other words, affect is the thing that is materially produced as always-already-social bodies relate to each other. Second, *feelings* on the other hand are the "proprioceptive and visceral shifts" in the background habits and postures of the body, which act as the body's instantaneous and situational assessment of a particular affect (Anderson 2006, 736). For Anderson, in other words, feelings refer to a particular body's personal, internal experience of a relationally produced affect. And, lastly, *emotions* are what we name in trying to describe the way that affect comes to be experienced through intimate, distinctly personal ways of being (Anderson 2006, 737), or in other words, the narration of feeling.

Significantly, in Anderson's conception, each of these three modalities are not independent but relational, and the sequence itself is not linear. What is important about these three definitions is that they allow us to recognize that the labels we use to *describe* certain feelings are not in fact naming universal sensations but rather contextualized, inter-body moments. This is not to say that people cannot appreciate, relate to, or even sense each other's experiences, for indeed it is possible to have empathy, but we cannot assume to understand or know the experiences of others *completely*. Further, we cannot assume that a particular social event or activity will produce the same sensations in *every* body. Fear, disgust, or pride, for example, are words that attempt to describe and categorize what are in reality thoroughly contextual and ever-changing sets of interdependent circumstances, which differ from body to body, space to space, and time to time. While these categories are certainly important, then, they are not the only matter.

If we want to understand the body as a social process and political tool, it is therefore important *not to stop* at these narrations. We cannot take the emotional label as an end in itself, but instead must look at what these named sensations seem to do, motivate, or promote, in terms of social (and individual) change. I want to suggest that the term "visceral" can help us to accomplish this goal, and to do so in a way that is perhaps easier to understand and relay than the language of affect, feeling, and emotion. To begin, it is obvious that visceral reactions lead to other events, whether everyday or historical: a saddened adult listens to a certain music album, a disgusted student rejects a plate of food, an energized voter casts a ballot for a particular candidate. Thus, it is easy to talk about the visceral in a way that assumes interest in the processes and shifts of viscerality, rather than just the names.

In addition, to speak of visceral (re)actions is to simultaneously refer to the relationality of affect, the textuality of emotion, and the experiential qualities of feeling. It is easy to illustrate, as I hope to, that despite a sense of "pure" naturalness, visceral reactions are a thoroughly social phenomenon—events that are produced out of the discursive and material relationships of socio-material life. As a broader term, it encompasses all of the important distinctions and nuances that affect and emotion scholars have articulated.

And finally, it is also quite easy to explain that visceral reactions can change: that we can become (more) conscious of our habits (i.e. our visceral tendencies) to react in certain ways, as they have developed in their broader contexts, and that we can work in various ways to begin to understand and promote the intellectual spaces and material opportunities that are needed to feel out new ways of being (i.e. our visceral latencies). In this sense, the language of viscerality allows us to understand the political significance of a body within a SGCP classroom, and the way in which this body can be used as both a method of social analysis and a tool for political action. At the same time, because this visceral process is not "clear cut" and simply "rational," these acknowledgments keep us open to be able to anticipate and imagine, both individually and collectively, the possibility of new visceral futures. In the next chapter, then, I turn to answer the question of how to do research on and with the body—that is, how to do visceral research.

2   Doing visceral research

Relationality is where knowledge is created, mediated, and ruptured, presenting itself for future relational events.

(Springgay 2008, 31)

We need to question the all-too-common assumption that there is one researcher, with an unchanging and knowable identity, and one project, with a singular unwavering aim.

(Crang 2002, 652)

Doing research on visceral matters is not exactly a straightforward process. After all, if the data you seek to "collect" are ultimately "non-cognitive, and in large part nonverbal, how can they be included in your research?" (Latham 2003). Though qualitative methods have gained considerable ground in geographic and social scientific research in the past decades, and have been increasingly acknowledged as acceptable "scientific" practice, qualitative research on/with the body now threatens to push the established boundaries of academic inquiry even further (Crang 2002). Nigel Thrift (2000), for example, laments the weddedness of qualitative researchers to traditional ethnographic procedures (interviews and observation), and calls for much more creativity in research methods if we are to broaden the range of sensate life we register (Latham 2003). Others propose that it is not so much the methods we choose, but what and why we assume we are "collecting" that needs to change. Gail Davies and Claire Dwyer (2007) argue for a rejection of the idea that the purpose of social science research is to generate clarity and precision, or to reduce uncertainty and ambiguity, suggesting that we need to revise our understandings of what social science research actually achieves (2007, 258). Along these lines, Mike Crang (2005) notes that, "qualitative research, despite talking about the body and emotions, frames its enterprise in a particular way that tends to disallow other forms of [emotional and bodied] knowledge" (2005, 230). In many ways, research on visceral matters requires that we approach our traditional research method/ologies with a new sense of dynamic, creative practice (Latham 2003).

But this is only half the matter of researching the visceral, for such research is not just *on* bodies but ultimately *with* bodies as well. As researchers we are motivated viscerally to engage in the work that we do (Bennett 2004). We go into the "field" (a fuzzy concept itself) already with a particular emotional attachment to our work— a visceral understanding of why it is important to do what we do. And we *do* our research with and through our emotionally articulated bodies, negotiating our way through various "field" experiences and (re)acting to the different events and people that we encounter. For example, researchers trained in feminist methodologies that seek to disrupt hierarchy in the research relationship often feel responsibility towards our research participants, and empathy towards those in oppressed positions (Naples 2003). We also can experience outrage, anger, hostility, or disgust if our research participants are engaged in oppressive acts, and thus we can have a sense of confusion over what to do or how to act towards such persons in light of a feminist training that teaches us to always be cognizant of the uneven power dynamics that are crafted within research relationships, often with researchers at the top (Kezar 2003). In these ways and many others, we use our (always minded, developmental) bodies as "instruments of research," (Longhurst et al. 2008) and yet we rarely consider how exactly we do this, or how this affects our data "collection," our "findings," and our subsequent "analysis" (Crang 2003).

This chapter reflects upon what it meant for me to do visceral research within the context of SGCPs. Over the course of my "fieldwork" experience, I came to understand this research, like the (un)motivated bodies that I wanted to "get to know," as ultimately relational in nature. As the chapter's opening quote by Stephanie Springgay suggests, it is in the relation between people/bodies/or things that knowledge is both created and disrupted. In this sense, the process of "getting to know bodies" in my research necessarily included my own (motivated) body, and it was through the interactions that I had with other bodies that I came to really "know" anything at all. This, of course, dramatically calls into question notions of "scientific objectivity," in which the collector of data is meant to remain distant from the objects of his research (a critique that feminist researchers have been advancing for years; see Haraway 1988). And yet it also does much more than this. It also calls into question how and why we (come to) know, as well as what we can (rightfully) say that we know, during the course of a research endeavor. Research on/with bodies within SGCPs demanded that I research as a practice of daily life—that I *not* separate what I was "looking for" from what I was living and doing. While I did engage in traditional ethnographic methods, including interviews and participant observation, I did not divide these research moments from other developmental moments of my physical and emotional life. What's more, I came to accept and enjoy, rather than explain away, the complexities and contradictions of this daily research life, seeking in the end not a single unified picture of SGCPs but rather partial descriptions of the actual, as well as partial suggestions of the possible.

This approach differs significantly from much of the published research that has intended to evaluate healthy eating/alternative food initiatives like SGCPs. Most evaluative research relies on methods of assessment that ultimately frame the success or failure of such programs in terms of their ability to produce a fixed set of results (Morris and Zidenberg-Cherr 2002, Morris et al. 2002, Veuglers and Fitzgerald 2005). For example, nutrition scholars Jennifer L. Morris and Sherri Zidenberg-Cherr (2002) wanted to test whether or not nutrition lessons combined with planting and harvesting would increase students' willingness to try new vegetables. They used questionnaires to assess students' knowledge of nutrition, and also a vegetable preference survey (which involved sampling six different vegetables) to evaluate students' openness to new foods. Their results suggest that "exposure to the [gardening and nutrition] curriculum *improved students' preferences* for several vegetables ... [and that] most of these *improvements* were retained 6 months after the completion of the lesson" (2002, 93; emphasis added). While this type of research is helpful for understanding general trends in students' food habits in relation to SGCP-like initiatives, and while the results ultimately indicate that garden experiences do affect some sort of change, such studies are predicated on the dubious assumptions that there is one right way to know food, and that this way of knowing food is both fully testable and knowable by "external" researcher(s). These assumptions, and the methods that they invoke, are not respectful of the social practices and the political contexts through which everyday life practices, like eating, actually unfold (Latham 2003).

My research begins with a different set of assumptions. First, it demands that we look at the broader context through which food preferences develop, and out of which food "choices" are made. This means both that we have to be aware of the different experiences and connections that students bring into SGCP eating events, and that we must consider how issues of access (in a networked, visceral sense) can affect the potential impacts of any "healthy" food experience (and thus the overall "success" of the program). A preference survey or standardized questionnaire does not allow for such depth. Second, my research approach demands that we understand eating as relational activity, bound up in the webbed and often quite haphazard unfolding of daily life. Seemingly mundane or chance interactions with peers, teachers, and (even) researchers thus literally *matter*, and cannot/should not be "controlled for" or written out of food studies as nutrition/health researchers often attempt to do (Perez-Rodrigo and Aranceta 2001). Third, I also recognize that we must consider how knowledge about what foods are "good" to eat, or what behaviors are said to constitute an "improvement," are not in fact truths from which we can assess SGCP achievements but rather socio-material constructions that themselves require further assessment. Thus, for me, evaluating the "success" of SGCPs cannot involve predetermined tools of appraisal like multichoice questionnaires or tastings of particular food but rather must be accomplished as a situated, co-creative, and dynamic process of reflection and (re)action by the participants themselves (students, parents, teachers, leaders and whole communities). Indeed, the researcher alone cannot effectively evaluate improvement or success.

It is important to point out that the shortcomings that are redressed in the above paragraph through my research approach are not limited to quantitative analyses of eating events. In fact, geographer Alan Latham (2003) has put forth a very similar critique of ethnographic research—arguing that qualitative researchers, too, need to do a better job of respecting the social practices through which everyday life occurs. Latham's call is worth reviewing at length because of how similar it is to what I describe above. His also has three parts: first, he says, we "must recognize that much social practice is different (but certainly not inferior) to the more contemplative academic modes of being in the world" (2003, 1998). This means that as ethnographic researchers, we cannot expect our observations of material life to fall neatly into the discursive categories through which we seek to explain it, and neither can we expect our interviewees to readily articulate or reflect on their own life practices through such practical and cognitive means. Second, Latham suggests that our research "must contain a sense that practices (and thus subjectivities and agencies of which they are a part) are shot through with creativity and possibility (even though these are "constrained" and limited by existing networks of association)" (2003, 1998). In other words, we must recognize that there is a latent, material agency in all socio-material interrelationships that exudes potentiality even as it is constrained by the structured tendencies of our uneven world. And last, Latham suggests that, "the everyday should not be viewed as a world apart from the more rationally grounded realms of social action such as 'the state,' 'the economic,' 'the political,' or whatever" (2003, 1998). In this sense, we also need to approach everyday practices like eating as processes in and through which broader political economic forces take shape and are constituted (as in Probyn 2000).

While the specific language of our methodological frames differ, both Latham's call and my methodology revolve around the same three central axes of analysis: the production of knowledge, the structure of power/hierarchy, and the unstructured ontology of lived experience/social practice. My three main research questions as I approached this study were also meant to address these axes:

1. How do the case study SGCPs reinforce or resist notions of expert knowledge in regard to what (and who) constitutes "healthy" "alternative" food?
2. In what ways do the SGCPs heighten or disrupt current patterns of geographic, economic, and cultural availability of fresh fruits and vegetables?
3. Through what means do the SGCPs provide students with opportunities to reify or reconfigure the boundaries of social identity and social practice?

Of course, while the above questions address knowledge production, structural forces, and networked or relational ontology, it is also important to note that each question itself contains the potential for both fixity and fluidity, and no line of inquiry on its own is complete without being complicated by the other two. In this way, the methodology that I use for this SGCP research represents a "rejection of singularity" as the objective of social science research (Davies and Dwyer 2007), and while it does not fully denounce the social science work of delineating tendencies or trends, it pushes this work towards addressing the realities of "everyday social practice in the articulation of these tendencies" (Latham 2003). The outcome or findings of my methodological approach therefore do not accomplish what typical social science research seeks or claims to accomplish: that is, a unified, logical distillation of the phenomenon at hand that is confirmable (by triangulation) and replicable (by future researchers). Instead, these questions have allowed me to pull together a description that is, like many feminists and qualitative researchers admit (Haraway 1988, Tsing 1993, Naples 2003), both partial and situated in the moment(s) of my research.

Yet my research also goes further than this. Although my "findings" say more about the contradictions and complexities of specific events of visceral relating than they do about how we can predict specific visceral (re)actions or generalize about visceral trends of (re)acting, they are accounts that are infused with a sort of "fidelity to what they describe" (Latham 2003). Following a core of human geographers whose research takes on a practiced/performative approach (Thrift 2000, Dewsbury 2003, Latham and Conradson 2003), I see description as allowing for a more subtle and loyal interpretation or imagining of how everyday life practices like eating actually unfold. In describing such practices, however, I do not want to imply an interest only in the intimate and the personal (Latham 2003). And indeed, I do not see descriptions themselves as *enough*, or as being already *in themselves* political. Instead, I want to use such descriptive imagining to illustrate the need to allow the complex and contradictory visceral realities of everyday life to imbue our politics, inform our policies, and (re)invent our pedagogies. It is in this way that such description can "go beyond" the limited case study boundaries of being "moments-in-time-and-space" and become something that, as we viscerally (re)imagine and (re)practice, can affect other political moments and mobilizations. It is also in this way that the expert-based knowing that is so typical of most research projects can become replaced with a diverse set of knowledges that do not privilege one way of (scientific, intellectual) knowing as *the* right way.

In this sense, I draw from a line of research ideas and practices that have urged us to study or research "with" (Pratt 2004, Sundberg 2004), whereby a researcher's "objects of study" become instead "speaking subjects" whose words and ideas are central to the production of research knowledge and thus can directly affect social change (Sundberg 2004, 271). I also draw from researchers like Ben Anderson (2005) and Springgay (2008) who take this task of "researching with" into the affective or visceral realm, insisting that speaking is just one means by which bodies and minds come to know and communicate knowledge. In this sense, the task of "researching with" also requires a certain attention to the affective processes of interaction between material bodies (including those of researcher and researched), recognizing that "listening with" (Anderson 2005) or "being with" (Springgay 2008) always involves more than linguistic communication, and thus that knowledge production in a research endeavor always involves more than cognitive thought. As Anderson's work illustrates, "researching with" as a visceral method is therefore also a means by which a researcher and his/her research participants may begin to collectively experience, discuss and make sense of visceral reactions; in researching with, we can begin to understand the ways in which our visceral judgments (in regard to food, music, ideas, etc.) are variously articulated through the material networks in which we are embedded. In so doing, we may also begin to relationally explore and feel out other ways of being/feeling.

Gardening and cooking with

The first method that I want to discuss in relation to my SGCP research is participant observation, or what I came to call a relational activity of "gardening and cooking with." This activity was ongoing during my 3 months at each research site, as well as my one-week initial visits to the schools, and involved participation in both gardening and cooking events, sometimes culminating in eating food with others. I recorded these events in a field journal, often jotting down notes during the events themselves, and then recalling and rewriting them in greater detail later in the evening. I also carried a tape recorder with me, into which I sometimes spoke verbal memos. All of these journal entries and memos were transcribed, reviewed for thematic patterns, and coded with labels that spoke to my three main axes of questioning: the labels included expert knowledge, peer knowledge, knowledge as power, economic inequity, geographic distance, racial boundaries, gender roles, class culture, pedagogical rigidity and redundancy, and flexibility/play in the class-room. Notably missing from this list are the perhaps more obvious themes of alternative food initiatives, including bodily health, environmental sustainability, and community development. These themes are not absent from this data but rather are reframed through the use of these other labels in a way that allows me to explore important and unanswered questions about alternative food: e.g. how knowledge about healthy eating is produced, who identifies with different types of local and non-local food products, and what practices and pedagogies effect change at the community level and beyond.

Plainville School, Nova Scotia

The specifics of gardening and cooking at each school site differed in some important ways, but the overall tasks were similar at each school site (see Table 2.1). Participant observation at Plainville School in Nova Scotia ran from the beginning of July 2007 to the beginning of October 2007, with an initial meeting in May of 2007. Although Plainville School was not in session until the end of August, the garden was alive with both people and plants when I arrived in July. Part of the reason for arriving in July was to experience the garden during the height of the growing season in Nova Scotia where, unlike California, planting and harvesting is limited to the warmer months of the year. For most of the months of July and August, I was engaged in weeding, thinning, and eventually harvesting the produce from the garden. In fact, for this time I was the primary caretaker of the garden—as was arranged by the head of the SGCP initiative. In the beginning, I spent many days tackling the overgrowth of weeds that had accumulated prior to my arrival, and getting used to the no-see-ems, mosquitoes, and other pesky insects that buzzed in my ears and bit my arms and legs. In the garden grew potatoes,

Table 2.1 Daily functioning of Central and Plainville SGCPs

	Plainville School (NS)	*Central School (CA)*
Garden		
Staff involvement	Coordinator plans garden, Teacher Garden Committee oversees connection to curriculum, Dora (kitchen staff) coordinates harvesting	One paid and two volunteer garden teachers coordinate garden, write lesson plans, and lead garden activities, other teachers use space when needed
Community involvement	Volunteers weed, harvest, and donate time and materials, much of community unaware of garden or SGCP initiative	Community members initiated garden project, community generally supportive of alternative food/local gardens
Student involvement	6th graders harvest for school lunches, k-6 choose seeds to plant and harvest, informal play kept to minimum	6th graders take gardening as required class, every other week for 90 minutes. 7th and 8th graders take cooking and gardening (both) as an elective
Kitchen		
Staff involvement	Dora (kitchen staff) coordinates school lunches and purchasing of food not grown	Two cooking teachers coordinate cooking classes with the assistance of a homeroom teacher
Community involvement	A chef volunteer comes to do special cooking activities with students	Various volunteers or student interns occasionally visit and/or help to coordinate cooking classroom
Student involvement	6th graders cycle through in groups of three to four for week-long cooking responsibilities (each group does this about four times)	6th graders take cooking as required class, every other week for 90 minutes. 7th and 8th graders take cooking and gardening (both) as a bi-weekly elective
Eating		
Staff involvement	Staff often eat the school lunches informally in the kitchen, standing and chatting to each other	Teachers eat the prepared foods with students, sitting around a table together
Community involvement	Parents, politicians, and other community members are occasionally invited to eat a meal with the students	Occasionally visitors and student interns will eat with the students if present during eating time
Student involvement	Students will eat the school lunches if they ordered them, or sometimes as a sample while cooking	Students end the cooking class by setting the table and collectively eating together, talking and sharing stories

Table 2.1 continued

	Plainville School (NS)	Central School (CA)
Cleaning		
Staff involvement	Kitchen staff (Dora) coordinates students cleaning	Staff are left to clean what students don't get to when the bell rings
Community involvement	Community members or visitors typically do not help clean	Community members or visitors typically do not help clean
Student involvement	Students are expected to clean and put away all dishes prior to leaving for recess	Students are expected to clean and put away all dishes prior to leaving the classroom (though they do not always finish this)

tomatoes, cucumbers, zucchini, green peas, pumpkins, peppers, corn, carrots, lettuce, and a variety of herbs: basil, dill, and coriander.

These initial days and weeks were in many ways primarily physical—involving little cognitive analysis or intellectual discussion and much body power. My journal entries reflect this, commenting on "the huge mound of dirt that is covered with lambs quarters and those deep-set vines" (Field notes, July 2007, NS). However, as I came to find my routine in the garden (and as the weeds became increasingly under control), I started to recognize how my minded body moved through the garden, experiencing it in ways that were relationally my own. For example, I recognized that my most physical days of sweat and toil in the garden were somehow comforting to me—comforting both because they partially relieved a sense of guilt that I had been carrying with me, worried about my inability to "give back" to the communities I studied, and comforting also because in the frustration and exertion of tending to the garden, I came to partially comprehend what my researcher-mind was grasping to know—how the garden might be experienced as simultaneously exciting and mundane, energizing and depleting. As the summer rolled along, I came to appreciate the various opportunities that the garden afforded me, finding creative ways to prepare a never-ending over-supply of zucchini, or swapping stories with other Nova Scotia gardeners about the struggles and joys of tending a garden.

Though much of my initial relating in Nova Scotia was between the plants and me, while tending the garden over the course of the summer I also ended up interacting with a number of people. The school ran a summer program, for example, so kids were frequently playing in or around the garden area, and occasionally, as the produce was ready to pick, harvesting zucchini or tomatoes to bring home or to eat on the spot. Parents and teachers also frequently walked by the garden, striking up conversations with the researcher "from away," or (more commonly) agreeably answering my questions as I pestered them for information or interviewing opportunities. There was also a much-discussed vandalizing incident, in which the shed door (always locked when I was not around) was pelted and smeared with green tomatoes. As the summer rolled on and the school semester began, my duties changed from gardener to cooking assistant, and most of my time was spent in the school kitchen, where 6th grade students, in groups of three or four at a time, would be pulled out of their regular class activities for a week in order to cooperatively run the school lunch program. I would accompany students out to the garden where we would harvest potatoes, green beans or whatever the recipe of the day called for. We would wash, chop and cook the vegetables, adding them to other store-bought items in order to make soups, spaghetti sauce, pizza or roasted potato wedges. Sometimes we sampled along the way, as we were urged to do by the main kitchen staff person, Ms. Dora.

We cooked to fill school lunch orders, which were collected at the beginning of each week, amounting to about a third of the total school population. (The rest brought their own lunch, as is typical in Canadian schools, since unlike the US, Canada does not have a national tradition of government-subsidized school lunches). Any students who forgot lunch, or did not have lunch (whether for financial or other reasons) were always given something to eat by Ms. Dora. Teachers often ordered lunch too, and received bigger "adult-sized" portions (I was included in this group). When the meals were ready, the students and I would deliver the food to each classroom, since there was not a central cafeteria. Then, we would return to the kitchen and eat our own food—either what we brought, or what was cooked—negotiating over who sat on the three kitchen stools, and who had to stand to eat. After doing dishes (a generally unfavored activity), the students would go outside for recess, and I would sit down to write up the day's events in my journal. For me, the school day was over by around 1:30—the students went back to their classes, and I went on to other activities, or home to rest and reflect.

Central School, California

Participant observation in Central School in Berkeley, California, took a somewhat different form. I arrived at the school in October, over a month into the school year. Because the school was much bigger than Plainville School, and the students could not be actively involved in producing school lunches for the entire student population, the purpose of the school garden and kitchen was instead to teach students how to grow and cook healthy meals and snacks for themselves, which they would eat as a group at the end of their cooking class. This group snack was in addition to anything they might have during their lunch period, although often what was cooked in cooking class and what was served at lunch were similar recipes. Thus, the lunch program was connected to the SGCP classes, and the food served in the lunchroom was seen as providing a complement to what students were learning in their cooking and gardening classes.

My primary role in the school was that of a volunteer in the cooking and gardening classes. I developed a rapport with one of the cooking teachers right away, Tiffany, who was young and, like me, interested in holistic nutrition and medicine. Tiffany invited me to sit at her table during cooking class, and I gladly accepted. I spent the first days and weeks glued to her table, letting the daily routine of the cooking room sink in (see Table 2.1). Eventually, I began to float to other tables in the cooking room (there were three in total), interacting with the other (head) cooking teacher, Sally, as well as the various homeroom teachers that would accompany their students to cooking class. Most of my focus, however, was in interacting with the students themselves. I would help the students with whatever they seemed to need, but my tasks seemed to frequently involve taking over the chopping of onions when their eyes became red with tears, as well as cleaning up after them when they (conveniently) ran out of time.

There were two groups of students that would circulate into the cooking classroom: the 6th graders, who took (separately) "cooking" and "gardening" as weekly required 90-minute classes, and the 7th and 8th graders, who elected (or were elected into) "cooking and gardening" as a 90-minute, bi-weekly class. The 7th and 8th grade students were the ones who I got to know most, as they were in the cooking classroom more frequently. By the end of 3 months, I knew all of them by name, and knew some of their life stories in depth (I interviewed about a third of them). At each 7th and 8th class meeting, one of the three tables of 7th and 8th graders would be pulled (sometimes with tired groans and complaints about the weather) out into the garden, for 60 minutes of garden activities, after which they would return to the kitchen classroom to eat with the other students. About half of the time when a table was pulled out, I would accompany the students outside into the garden. This decision depended on where my volunteer labor seemed to be needed most, or sometimes where it seemed I would get in the way least.

The garden was a larger space than the Nova Scotia garden, and we grew a lot more variety. We had some vegetables that one might expect: different types of tomatoes, green beans, lettuce, radishes, and peppers, garlic, kale, collard greens, and summer squash. We also had some more unusual items: lemon cucumbers (small round spiny cucumbers that have a lemon taste), and figs (which were harvested very occasionally). Harvesting rules were a bit more relaxed than in Nova Scotia, and students frequently came out to the garden to pick and eat tomatoes, radishes or cucumbers during their lunch break (maybe due to the fact that Central School was a middle school, and unlike Plainville, did not have young children who might need more supervision in the garden). There were several different areas in the garden for sitting and relaxing, including a colorful bench. The Central School garden also had a chicken coop, with chickens that all had names and eggs that we would use in the kitchen.

I felt less comfortable in the garden than in the kitchen of Central School, particularly because it was less clear what I could do to help. It is not that I did not know how to garden, but in contrast to Nova Scotia, I was not in control of this garden, and I didn't know its rules of operation. The cooking room at Central School appealed to me more because I felt like I could figure out how to fit in and what to do in order to be of use; in the Central School garden, it was unclear how I fitted in and what exactly I could do. There were many awkward moments in the garden. Like the students, I would stand around and wait for a prompt from the garden staff—a staff of three including one full-time employee, and two volunteers. We forked beds, pulled weeds, sifted compost, planted cover crop, harvested produce, and sometimes played games like garden jeopardy. If I didn't know what to do, I got a shovel or a rake and made it look like I was doing something. At the end of the class period, I would return with the students to the cooking room to sit and eat with the students. The garden staff usually remained behind in the garden, and often did not eat with us.

Ontological messiness

In the process of "gardening and cooking with" in Nova Scotia and Berkeley I interacted with a lot of different bodies. These bodies were not only human bodies—students, teachers, and volunteers—but nonhuman bodies—insects, weeds, shovels, rakes, and a variety of different raw and prepared foods. I related to these bodies, and they related to each other, in and through a variety of verbal and nonverbal forms of interaction—telling stories, sensing and describing sensation, smiling, resting, sharing tasks, asking questions, digesting, appearing unsure, slouching, feeling annoyed, being hungry, following orders, and so on. All of these forms of interaction made me aware of the relationality of my own existence, as well as that of those around me. The data that I collected from these interactions describe such relationality with a sort of "fidelity" to ontological messiness (Latham 2003). They describe the randomness through which both chance and nonrandom factors and conditions come together in people's lived experiences of the SGCP initiatives. This fidelity is not to suggest that all forces or conditions are equally relevant or meaningful, as my interview data further specify, but it is meant to emphasize the haphazard contexts in which eating decisions are made. For example, consider the following field journal entry from a day in Plainville's SGCP:

Field Notes, Sept 2007, NS

Tina and Nick and Jim were in the kitchen with us, and the boys were really making a stink about the hairnets (taking them off whenever they could, and making a big deal about how itchy their heads felt with them, etc.). Tina didn't say a thing about it. We all went out to the garden to harvest. Nick really got into picking the carrots and so we picked all the rest, because there weren't many, but the best of the bunch were the ones we picked today (no worms). Tina and Ms. Dora picked a huge turnip, and I gave Jim the job of picking beans again. We also got some onion and Tina did the potatoes again, plunging her hands way down into the soil like she did last time. We got a lot of stuff, and used most of it in the vegetable soup. Ms. Dora used canned tomato and store-bought celery but all the rest from the garden, beans, onion, which was all blended to "hide it" from the other students (Tim didn't like this idea at all, and said so), and then potato, carrot, turnip. I noticed while we were busy cooking that the boys and Tina, too, really looked for things to do. I mean, they were happy being in the kitchen and when one task was finished, they went on to the next. Nick even asked me if he could scrub the rest of the potatoes! They really took it upon themselves to figure out what needed to be done. There weren't too many directives given by Ms. Dora . . . they got the swing of things, and for the most part were able to do it themselves. Ms. Dora said all the kids who cook with her had to try the food because it is an important part of the cooking process, learning what the food might need in terms of spices and salt and stuff. I made Nick a special bowl with only potatoes, carrots and broth, because he is the class's "picky eater," an identity he seems to enjoy (in terms of the attention it attracts). Tina liked [the soup] but doesn't like tomatoes, so she also ate a different lunch. All the teachers loved it, and the principal, who came for seconds. The good smell seemed to attract a crowd of staff, and the kitchen became (as usual) the hang- out spot for staff, with chit-chatting, some of them talking to me, and teachers coming and going.

I quote my field notes at length here to give the reader a sense of both a typical day in a SGCP classroom, and also a typical field journal entry. In regard to how such an entry would speak to my research questions, there are a few points in this entry that stand out. The authority figure in this case is clearly Ms. Dora, and secondarily me (since Nick asked me for permission), but the mood of the kitchen overall did not reflect an authoritarian space or a strict set of rules; instead, it was easy going and pragmatic, with everyone contributing, and fun, informal interactions certainly "allowed." Tina, who often was most keen about eating whatever we cooked (partly because of the financial situation of her parents, and thus a lack of anything else to eat), decided that she would eat something brought from home that day, because she didn't care for tomatoes. Nick, the designated picky eater, was very put off by the idea of blending vegetables to hide them in the soup, and yet when I fixed a special bowl for him with only potatoes and carrots, he gladly tried some. There was a drama over hairnets between the boys, which Tina took no part in, though this was a fairly typical ritual in which many students engaged; hairnets were considered "dorky" and carrying on about them perhaps a sign of "coolness," a hierarchy that Tina seemed to not want to be a part of. The dorkiness was most likely a result of various identity struggles that are repeated frequently in my data, including the association between hairnets and working-class food venues, and the gendered coding of hairnets as related particularly to women's work in such food service jobs—e.g. the lunch lady. (The counter to these hairnets, costume-wise, would perhaps be the chef's hat, which is often coded as both masculine and socially elite.)

The point of such recounting, however, is not to try to explain the whys of particular actions: Tina's lunch choice, the boy's (re)action to hairnets, Nick's acceptance of a bowl of potatoes. Rather, the point is to emphasize that because eating decisions are always made from within the haphazard but not fully random web of life forces, the whys are often as emergent as the what ifs. There are hierarchical structures that inform the above eating event, lending Nick an opportunity to get attention from an authority figure, and Tina a hungry belly with which to crave food. And yet there are disruptions and opportunities—a researcher in the room, a sense of hanging out, a wafting odor—that render particular outcomes uncertain and particular lessons unclear. While the whole event was conducted under the auspices of a "healthy" lunch initiative, the complexities and contradictions of the (re)actions and behaviors also defy definitive labeling.

Talking with

Beyond participant observation, interviewing was my primary means of data collection for this SGCP research. Similar to my participant observations, I want to frame my interview sessions as instances of "talking with," rather than one-way question and answer sessions. There are many reasons for this. First, my interviews ranged from semi-structured in some cases to open-ended in others. I often did not follow a specific interview script because I wanted to dialogue with people about what was important to *them* about food, rather than allowing my interests and concerns to dominate the discussion. This follows from my desire to undo nutrition intervention as a de-contextualized and universal practice (Hayes-Conroy and Hayes-Conroy 2013). While most of my interviews were audio taped, a lot of the dialogues that I had with people were also accomplished as part of a broader and more social event—usually involving eating or drinking—in which questions and answers were interspersed with other activities. To be sure, this is not to say that there was no perceptible hierarchy in the interview relationships. Especially in my interviews with students, but also frequently with SGCP parents, my status as a "researcher" carried with it an air of expertise that led many to assume (at least initially) that I was there to test their knowledge, rather than learn from them or with them. I tried to dispel this assumption through how I framed my questions and my research motivations over all (explaining, "I want to understand what the experience is like for you"), and also through how I conducted the interviews (offering information about my own life and motivations, sharing a meal or a snack, and inviting students to dialogue with their friends in addition to me). In particular, almost all of the interviews with students were group discussions, with food (not usually or all "healthy") as a centerpiece. Even so, there were occasional moments where I needed to directly address this perceived "expert" identity, such as when a student posed the following question during a group interview, before deciding between a donut and an apple slice (both food items that I had brought for the group):

> *Student*: Are you writing down what food we choose?
>
> > (Student, CA)

> *JHC*: No. I am not here to test you on your food choices.
> [she then took one of each]

Another reason why I want to consider my interviews as instances of "talking with" is because in conducting interviews, I frequently shared with participants my own interpretations and analyses of the SGCPs, and also my own theoretical background (e.g. my interest in issues of social difference and diversity, my worries about economic and cultural access, and even my attention to how visceral judgments are made). This two-way interview style is similar (but perhaps not as theoretically revealing) to Geraldine Pratt's project in *Working Feminism* (2004), a project in which the author engages her research participants in conversations about social theory as a practice of bringing theory to bear on the research process itself. It is also similar to the concept that an interview relationship can be a potentially "transformational" one (Kezar 2003), one in which the objective is not (only) to gather information, but to collectively produce new understandings that might directly impact both the "writing up" of the research and also the matter of the research itself (in this case, the functioning of the SGCPs). In this sense, I engaged my interviewees—all of them to some extent, but especially the teachers/volunteers and activist/leaders—in a process of collectively thinking through how the SGCPs work, and also how they might work better (that is, in more progressive, non-hierarchical ways).

A third way that my interviews became instances of "talking with" is through direct attention to the visceral body, and moreover, to the relational connection between bodies. I mean this in two ways. First, as I described in the introduction to this chapter, I was aware of my own visceral reactions and motivational drives as I engaged in the research process. This included often-conflicting feelings like guilt, anger, responsibility, worry, and excitement in response to what others were saying to me (even over the course of one interview). This also included sensing and responding to the various nonverbal cues of others: shifting uncomfortably, changing the tone of one's voice, smiling, appearing bored or animated, etc. I often took note of these visceral interactions in my field journal by way of a comment on the overall "mood(s)" of the interview. These relationally produced visceral (re)actions are important in that they propelled me during the interview process, compelling me to dig further into certain questions when it seemed appropriate to do so, or to back off and try a different approach if the dialogue was feeling unproductive. They also motivated me in terms of whose opinions I sought out overall (e.g. wanting to talk with food activists in the broader community, parents from diverse backgrounds, student voices). At various points in the chapters that follow, I directly discuss these visceral moments in explaining how specific questions were asked and answered.

The second meaning of "direct attention" to the visceral body is in the sense of an intellectual attempt to bring discussions and theories of the body into the interview discussion itself. It is in this sense that Anderson (2005) uses the term "listening with," a method whereby he listens to music alongside his interviewees and engages them in dialogue about the visceral sensations that drive and comprise the listening event. Similarly, many of my own interviews involved attempts to directly "get at" what the SGCPs felt like to the different people towards whom they were aimed to "touch" (Springgay 2008), especially students and parents. In other words, I questioned how certain food/events were viscerally appealing/animating to some and repulsive/chilling to others. It is important to note, however, that in following this line of questioning I quickly became aware that I could not expect my interviewees to have coherent answers to these questions, nor could I expect their answers to remain consistent over time. While this was at first a frustrating experience, as Latham notes of his own dilemmas with explaining social practice, making sense of why this line of questioning is particularly difficult to interviewees is part of what it means to take the "flow of practice and its complex embodied subjectivities seriously" (Latham 2003, 2000). In asking myself why these questions seemed to put people (especially children) on the spot, and why I initially expected my participants to be able to explain and make sense of their actions, I became increasingly aware that it was the difficulty of articulation, and the shifting complexity of the answers, that best explained what I wanted to know. With this realization, my questioning shifted from "why" questions (which seemed to take food out of context) to "how" questions (which asked about specific events):

> *JHC*: What is your favorite thing you make?
> *Student*: Um, pizza.
> *JHC*: Why do you like it?
> *Student*: I don't know, it just tastes good, I just like the taste.
>
> (Student, NS)

> *JHC*: Would you make any changes to the program if you could?
> *Student*: More meat, I mean not too much, just a little more, and less on the vegetables. I like pepperoni pizza, but we aren't allowed to have it. Stuff at school is healthier than at home, like we like to eat chips at home.
> *JHC*: How does it feel to have different food at school and home?
> *Student*: Sometimes it feels weird because when I am cooking at home I go for stuff we have at school and we don't have it at home, so I am like, what am I gonna put in this now?
>
> (Student, CA)

Over the course of this SGCP research, I conducted interviews with 144 people: 63 in Nova Scotia, and 81 in the Bay Area of California (see Table 2.2). I divide my interviewees into four general types, in regard to their association with the SGCPs:

1. teachers/volunteers, a largely female-identified group who are engaged in the daily running of garden and/or kitchen classrooms;
2. parents/guardians, a group of mostly mothers whose children attend one of the case study SGCPs and who may or may not also engage in volunteer activities in the classrooms;
3. current/former students, a diverse group aged 11–14 who either are currently enrolled in a SGCP class or have been enrolled in the past 2 years (as of 2007); and

4. leaders/activists, a diverse group who are engaged in either promoting and lending ideological influence to SGCP-type projects broadly (or the SGCP projects that I studied particularly). These are "leaders" who are engaged in promoting/pushing the boundaries of alternative food in other ways (for example, in ways that critique alternative food's "whiteness" or "eliteness," or that provide alternative food opportunities for more people).

Some, but not all, of the leaders and activists that I spoke with included those associated with the Slow Food movement (which, as I mentioned, was the result of a parallel project headed by my colleague/sister, Allison Hayes-Conroy).

Table 2.2 Number of interviews by type

Types	Nova Scotia	California
Teachers/volunteers	7	10
Parents/guardians	27	24
Current/former students	15	29
Leaders/activists	14	18
TOTAL	63	81

These interviews lasted anywhere from 15 minutes to 3 hours, though many fell in the range of 30–60 minutes. The interviews with students and parents were typically shorter due to time constraints (most were accomplished during or directly after school hours). Interviews with teachers/volunteers and leaders/activists were often longer. All interviews with students were either conducted in a peer group setting, or in connection to an interview with a parent. Most interviews were tape recorded and transcribed verbatim, but others were recorded through written notes either by myself or by a research partner (Allison). A few interviews with parents were conducted over the phone, when a suitable meeting time could not be found, and two interviews were conducted by way of a series of email exchanges between two parents and myself. All of the transcripts from these interviews were subsequently coded and analyzed using the same thematic labels that I used in analyzing my field journal notes. To enable the reader to distinguish between field notes and interviews, all large excerpts from field notes in this book are presented within a framed text box (short quotations will be denoted with a parenthetical note), while all interview data will either be distinguished by quotation marks or, where several responses are recorded from a group, the words of interviewees will be displayed as indented text. In both cases the quotation will be followed by a label: e.g. (Teacher, CA). While I often avoid using names altogether (because the quantity of interviews would make names superfluous), any of the frequently repeated names that I do use within this document—with the exception of public figures—are pseudonyms. At times, I also deliberately change the wording (but not meaning) of quotations so as to not reveal other potentially identifying information.

Overall, my interview data speak strongly to all three of my central research questions. Often within the course of one interview, my discussions with research participants would reveal stories about how food knowledge is produced, policed and reposited, how opportunities for interacting with and experiencing certain foods/events are distributed unevenly across the social terrain, and how engagement with food in SGCPs is an emergent and unpredictable process.

Researching with

Beyond participant observation and interviews, I engaged in a number of creative, participatory classroom activities with the SGCP students. These activities included peer interviewing, field journaling, and letter writing. Collectively, I call these activities "researching with" because they allowed me to accomplish a variety of important research tasks through direct inclusion of students in my research process—allowing that I, individually, was not alone in the task of conducting research. These activities began with visits to a number of "homeroom classrooms" of the students engaged in the SGCPs. Over the course of my research, I visited six classrooms of approximately 25 students each. In Plainville School, I conducted research activities in a Grade 2 class, a Grade 4/5 class, and a Grade 6 class (while only the Grade 6 students were involved in cooking, all of these classes had some engagement with the school garden and the lunch program overall). In Central School, I conducted research activities in three 6th-grade classes (which I sought out instead of 7th or 8th grade classes because I had already interacted with 7th and 8th graders through group interviews). These classes were also chosen out of convenience, since the homeroom teachers of all of these classes requested that I visit and speak to their students about my research activities.

Peer interviews and field journaling

In all of these classes, I conducted peer interviews and field journaling activities, tailoring the activities in ways that were appropriate to fit each age group. The peer interviewing and field journaling activities allowed me to introduce myself and my research to the students, and further, to teach them how to engage in the types of research activities that I was conducting at their schools. Peer interviewing began with a brief lesson about what research is in general, and what interviewing accomplishes as a method. After this lesson, each student was given a basic interview script that included several questions about their SGCP experiences: e.g. "Do you like the foods that you learn to make in the cooking class?" or, "Do you think it is hard to eat healthily?" The students were also encouraged to be "good listeners" and to respond by asking one or two follow-up questions of their own choosing. Some suggestions for follow up were also listed: e.g. "If not—why not? What would you like better?" or, "What would make it easier to eat healthily?" The students were instructed to write down notes about what their peers said to them, and then to switch interviewer/interviewee roles and repeat the process.

After the peer interviews were conducted, I introduced field journaling as a take home assignment, building from the interview lesson. I briefly introduced the concept of participant observation, and I instructed students to engage in such an activity during one of the next meals that they ate outside of school (at home, or in a restaurant). The students were to take notes on their meal using a note-taking guideline that encouraged them to describe the experience in depth: "Describe what you ate," "Describe the setting," "Describe who you ate with," "How long did the meal last?" "What was the mood of the meal?" These homework assignments were due the following week, and were collected by their homeroom teachers. Neither their homeroom teacher nor I graded the students on either assignment.

Letter writing

In addition to peer interviewing and field journaling, I conducted a third "researching with" activity—letter writing—with three of the same classrooms of students—two 6th grade classes from Central School, and one 6th grade class from Plainville School. These letter writing activities began with a lesson on how to write a letter, and were conducted collectively by the homeroom teacher and myself in Central School in December of 2007. Students were encouraged to think about what they wanted to know about the other SGCPs, and to brainstorm some ideas about what was important to tell the students about their own garden. I gave them some prompts in this regard by explaining a little bit of some of the differences between the Nova Scotia and California SGCPs—e.g. different growing seasons, knowledge about fruits like persimmons or pomegranates, cooking for classes versus the whole school, etc. The students then were given 30 minutes to write a letter to a particular class member from the other SGCP (names were distributed randomly). Some students took longer, and some even took the letter home to finish. These letters were then sent to Nova Scotia, where the Grade 6 teacher in Plainville conducted a similar lesson on letter writing. In March of 2008, response letters were sent back to Central School (via me in Pennsylvania).

Through different means, all three of these "researching with" activities allowed me to better understand how students talk about food in general, and their SGCPs in particular, within a peer group setting. In short, they gave me insight into what students themselves really find important, as opposed to what teachers or parents say through speculation and observation. Admittedly, the prompts for these activities came from adults. Yet, at the same time we purposefully encouraged a good amount of play within these activities, which led to a wide range of responses. For example, during the letter writing some students wrote relatively little about the garden or cooking classes at all, choosing to focus on other more seemingly pressing particulars of their life (other classes or interests, who their friends are, etc.). Also, the details that students chose to focus on in the field journaling were wide-ranging in both level of description and areas of concentration. One student explained: "I ate my dinner at a circular table, set for three, since mommy is in France,

Sammy ate on the floor, as a dog should. Our kitchen is at the back of the house, and dark. We had a candle lit." Another simply wrote: "I ate on the sofa, with crumbs everywhere!"

The data that resulted from these activities of "researching with" are incorporated into my research in several ways. Primarily they exist as background data, which came to inform the types of questions I asked during interviews, and also the ways that I articulated my questions (particularly during interviews with students and parents). These data also came to influence the ways that I asked teachers and leaders to think about the contexts in which students are motivated to eat, and also about how students' different backgrounds at home might affect their experiences at school. In addition, the actual *experience* of conducting these activities was itself something that I considered relevant as data, and thus something that I wrote about in my own field notes as well. For example, I was intrigued by the sometimes contentious dynamic that the peer interviewing seemed to set up between students who were not "accurately" recording each other's words, or who were not asking the "right" follow-up questions. At one point, one student said to another: "No, don't ask me that; ask me about [this]" (Field notes, September 2007, NS). Finally, I also occasionally include direct quotations from the homework or letter writing activities when I think that the data speak directly and succinctly to my research questions.

Living with

The last activity that I want to discuss is something that I broadly call "living with," recognizing that the relational reach of "fieldwork" goes well beyond the specific SGCP school contexts, and indeed calls into question the boundary of the "field" itself. In my view, this research activity includes everything from initially choosing the sites of my research itself, to traveling and cohabitating with others, to personally making food choices. Over the course of my 6 (plus) months of fieldwork, my data "collection"—the ways in which I came to know and understand what I did about the SGCPs—involved much more than participant observation, interviews, and classroom activities. It involved many other practices of life as well: shopping, cooking, exercising, listening to music, making and meeting friends, reading, exploring, taking the bus, helping others, being invited to parties, or asked to host them, and so on. While ethnographers who are keen on "full immersion" in one's research context might not find it surprising to include these activities as part of one's methods, I want to explain "living with" not as an immersion situation so much as a bodied, relational activity that implicates the researcher even more directly in the production of her own data. In this sense, "living with" is a method of its own particularly because it highlights how my own daily practices and habits of being came to influence my understandings and experiences of (researching) the SGCPs. Below I discuss three instances of "living with": choosing Nova Scotia and California, engaging with family and friends, and making food decisions.

Choosing Nova Scotia and California

In the introductory chapter, I previously discussed the decision to choose Nova Scotia and California as the two sites for my research; here I would like to specify how this decision was (in my definition) a particularly visceral one, based on both an intellectual understanding of alternative food organizing in each place, as well as a sense of curiosity and challenge in regard to what I did not yet know. The thought of living and researching in both places intrigued me, but the choice of Berkeley was perhaps an easier and more obvious one to make. The ESY project was written up continually in the media, often in ways that cited Alice Waters as a key figure of a SGCP movement or "The Delicious Revolution" (Waters and Berkeley 2006, Brown 2007). More generally, the Bay Area was frequently cited as a place where alternative food was *happening*, through a number of different farmers markets and urban community gardens (Severson 2008). Indeed, California is a unique and significant place to study alternative food (Guthman 2004). It is often considered by food scholars and activists (including many I met in Nova Scotia) to be the hotbed of alternative food activism in North America, from which a great deal of ideas and (organic) foods emerge. My impressions of Berkeley, based on my preliminary research and also discussions among activists and "foodies" (those who enjoy alternative/fancy food), were largely confirmed upon arrival. As several of my student interviewees explained of the area:

Berkeley is an oasis.

(Student, CA)

Berkeley has all of these hippies. It is because people care more here. It is a really green city, and it is a really good thing.

(Student, CA)

In contrast to Berkeley, Nova Scotia is, at first take, a more curious choice. Alternative food activism is not nearly as commonplace as in Berkeley, and attention to food in schools was just beginning to gain ground as I started my research in 2007 (CBC 2006). In addition, local food is notably harder to *do* in Nova Scotia, as the growing season is fairly short in the region and the variety of produce able to be grown is much smaller. Part of the reason for my curiosity over Nova Scotia, however, stemmed from such differences from Berkeley. Unlike the urban context of Berkeley, situated within a metropolitan area of over 6 million, the rural area surrounding Plainville School consists of small municipalities of less than 2,000 residents. The nearest major supermarket to the school is 30 minutes away, and the nearest farmers market (other than roadside stands) is 45 minutes. Halifax, the nearest major metropolitan center (population approximately 360,000) is an hour's drive. Although there are notable circles of "foodies" and alternative food activists (mostly centered in Halifax), I also sensed that Nova Scotia was often considered the opposite of Berkeley food-wise: a rural province that was somewhat "behind" the times in relation to the "cosmopolitan" food cultures of urban places like the Bay Area. Indeed, when I arrived in Nova Scotia, many of my interviewees described the food "scene" around Plainville in the following way: "Here, we don't really have any food traditions. It is rudimentary, simple" (Mother, NS).

In other ways, however, the two sites and programs are congruous. The Plainville SGCP was inspired largely by the ESY project and the ideals of Alice Waters, and it was supported in part by Nova Scotia's Slow Food chapter, a group of motivated alternative food enthusiasts who share an interest in local, sustainable, and heritage foods. Like a lot of the SGCP leaders in Nova Scotia, Alice Waters is, too, a member of the Slow Food movement; in fact, she is currently the vice president of Slow Food International. Similar to California, Nova Scotia had also received some recent media attention, due particularly to changes in school food policy passed at the provincial level (Nova Scotia 2006). Like California legislation (CA Gov 2005), these policies advocate a gradual but firm shift to "healthier" foods in schools. Nova Scotia has also been an area of intellectual interest in regard to food, particularly around issues of hunger and food security (Williams et al. 2006). In addition, both Plainville and Central Schools serve a diverse population of students: Plainville in terms of economic/social class differences and Central in terms of both class and racial identities. In both cases, program leaders and advocates argue that the SGCPs particularly serve disenfranchised populations.

This combination of ideological similarities (in regard to alternative and healthy food) and geographical/social differences (food culture, diversity, urbanity, etc.) compelled me to take on this SGCP research at both locations, particularly because I was curious about how the ideological similarities would be taken up and transformed by the particular contexts of the school sites. Nevertheless, although a comparative analysis might be appropriate to such research, I came to understand these sites not as opportunities to find clarity through differentiation but instead as occasions to "witness" and be a part of the making of food decisions as a contextual practice (Dewsbury 2003). The more that I witnessed food in these contexts the more I was shocked by the ideological claims of some alternative food leaders and nutrition experts, claiming to have all the answers. How could anyone possibly have all the answers for everyone, everywhere, about what is "good" or "right" to eat? The more I witnessed, the more I found such claims absurd. At the same time, however, I also came to understand the power of such ideological claims, and thus also their potential/partial usefulness in re-doing alternative food. What propelled me through this lived research, then, was both a visceral determination to *not* simplify food practice down to any ideological claim, and also a visceral curiosity to keep open the possibility of multiple, contradictory answers.

Engaging with friends and family

For me, one of the most memorable features of this research experience is the fact that I conducted this research alongside my colleague and twin sister, Allison, who was at the same time engaged in a similar research project on the Slow Food movement in Nova Scotia and Berkeley. Not only did we choose our research sites in (visceral) concert, but we also traveled and lived together during our fieldwork, and we often accompanied each other on research endeavors (although I followed her more than she did I due to issues of space and admission into the schools themselves). This dual effort is significant for many reasons, but three stand out in my mind. The first and perhaps most obvious is that in helping Allison with her research, which included accompanying her to many (about 60) of her interviews with Slow Food members and other food activists, as well as outings and other activities (about 15), I was able to gain great insight into the background of alternative food issues in each location. In addition, I was introduced to a variety of people that became informants in my own work; indeed as our research projects progressed, Allison and I conducted a number of collective interviews that were relevant to both her and my work. In seeking out and conducting these interviews, Allison and I also particularly became aware of the significance of looking like one another. The fascination that many people appeared to have over the joint (twin) effort of our research endeavors seemed to make rapport with research participants come easily, perhaps because the research was so noticeably enjoyable for us.

Second, researching with Allison is also significant from an intellectual sense in that she and I are interested in similar theoretical work—relational ontologies, feminist theory, affect and emotion. Thus, Allison and I were not only researching alongside one another, we were also (re)analyzing and (re)practicing our ideas and interpretations along the way. The impact of this is not minimal. For example, my research was largely motivated by feelings of responsibility towards those who I considered disenfranchised in/by the alternative food movement (those for whom the goodness of alternative food was not already a given), while Allison was meeting and dialoguing largely with the "other side," a group of alternative food activists that have been described as culturally homogeneous and economically elite (Leitch 2003). Seeing both sides of this alternative food struggle forced me to recognize the inconsistencies and contradictions of such polarizations, and also called me to question how anyone can really know or fully define the boundaries of progressive activism within alternative food. Similarly, my drive to understand the experience of alternative food through the lens of social difference compelled Allison to pose more questions to her research subjects about the homogeneity and (lack of) outreach in Slow Food, allowing me to witness (through her interview process) how my own research questions play out in her own work. This back and forth between our two projects was thus indispensable to how I came to understand and analyze the SGCPs.

Third, the circumstance of traveling and researching with Allison brought me into contact with a number of other influential people who were not just "research subjects" but who primarily became friends. These included two men who we lived with (at different times) during our stay in Nova Scotia, John and Paulo, and one woman that we lived with in California, Lisa. John is an older food activist and enthusiast, and someone who we shared a house with and cooked meals with for a month and a half while in Nova Scotia. He not only gave us entry into the alternative food world in Nova Scotia, but he also gave very generously to us in terms of his time and his space. Through getting to know John, we learned much about his own food-based motivations, but also much about our own work and our ways of thinking about/approaching alternative food. I came to respect John's opinions a great deal, and to see him as very knowledgeable and thoughtful in his approach to alternative food. At one point, Allison and I sat down to dinner with John with the tape recorder rolling (since our conversations often revolved around food and our research experiences, and there was never a moment when we really weren't *doing* research). Our conversation that night was pivotal to how I came to understand food decisions.

I was arguing with John about elitism in alternative food, suggesting that the "positive politics" model (avoiding negative confrontation) eschewed the fact that alternative food was already off-putting to some people. I gave him an example from my research: one girl whose working-class father came to a quite fancy SGCP event and told her, "I'm not gonna eat that [a salad she had helped make]; I'm not a rabbit" (Father, NS).

I implied to John that alternative food activists weren't paying attention to the visceral trauma that could come from what I read as a class-based clash between elite "foodies" and working-class parents. John responded to me:

> Let's just analyze what you said there for a second . . . A father who said, well, "this is rabbit food"; so does that mean the whole experience was a negative experience? Not necessarily . . . if you ask the kid in the presence of the father she might echo what the father said just because . . . but the father may have no understanding at all of what it was like for the kid to go out in the garden for the first time and harvest something out of it . . . whether it was lettuce a carrot, or the whole concept of harvesting something from the land and seeing bugs or dirt on it that you don't see at the grocery store, or maybe it was a crisp sunny day and just being outdoors was an elation, they weren't getting sitting in a classroom . . . the parent has no idea what those things were like in terms of an experience for the kid, they are just equating one aspect of the contact of that kid with the experience to their preconceived idea about what constitutes good food, so because that comment is there it doesn't mean the experience for the kid wasn't worthwhile overall. There are so many dimensions to an experience . . . you don't know what the impact is going to be, you can't know what part of that experience somebody is going to run with and make a decision later on about based on that experience . . . like that kid—the kid may never eat another salad because dad says its rabbit food, or 10 years from now she might grow a garden because the whole idea of going outside in the fresh air and picking something out of the ground resonated somehow.
>
> (John, Activist, NS)

I quote this response at length because it was a moment in which I realized that my intense feelings of responsibility towards any perceived "underdog" were actually thwarting any opportunity for reading agency into such situations. Both he and I were right to some extent—some of the alternative food leaders that I was talking to *were* largely unaware of, and even outright antagonistic towards, these situations of class-based social difference within the SGCP events. But still, this did not mean that the "outcomes" of the SGCP were predictably negative. To the contrary, the "outcomes" of the SGCP were highly unpredictable, at least in terms of influencing students' long-term motivational drives. I had (and still have) no idea whether this girl will choose to shop at a farmers market 10 years from now, but I do know that her experience of the SGCP event was multi-dimensional and complicated. Was that enough to come home with? Was that a research "finding"? John's answer to this was, well, regardless, that is the truth.

Over the course of my research there were many other epiphanic instances like this one, some of which I describe in the chapters that follow. Chapter 4, for example, tells a "tale of two dinners," a two-event evening to which my other friend Paulo accompanied me. While Paulo was neither a foodie/activist nor a local Nova Scotian, his presence at the events and his discussion with me following them gave me important insights into the community politics that surround Plainville's SGCP. Similarly, many of my interactions with Lisa, an alternative food/yoga enthusiast, and her young children, who both attended public school in Berkeley, influenced me in tremendous ways—particularly in rethinking the (non)centrality of leaders/ideologues like Alice Waters to the everyday playing out of SGCPs on the ground. These interactions with friends and family illustrate how the relational method of "living with" goes well beyond an ethnographic "immersion" situation. More than any other event or phenomenon in my research, these relationships called into question the boundary between researcher and participant in this SGCP research.

Making food decisions

The third instance of "living with" that I want to discuss is the matter of my own personal food decisions. In light of the subject matter of this research, I think it is important to discuss and reflect upon the way that I myself make food decisions —acknowledging that as a researcher, my visceral (re)actions towards food itself are also relevant to this SGCP study. To the extent that it is actually possible to "know" one's own visceral judgments, then, I will try to explain my own positioning on food.

As my conscious mind distinguishes, my food decisions often seem to fall into one of two categories: social decisions, involving not only food itself but also a broader context of other people eating and offering food, and personal decisions, where I am not very concerned with a broader social context and I am deciding what to eat based on my cravings, habits, and/or momentary needs. Socially, I will eat almost anything. Over the course of this research, I was offered and accepted food many times that under different circumstances I might not "choose" for myself (e.g. coffee, pork fat and sea urchin for breakfast). This is not to say that I often felt forced or obligated to eat food that I did not want or care for; on the contrary, I did usually *desire* to eat such food precisely because it was a shared experience, and because the sharing of food in social contexts is something that has always been particularly important to me. In this sense, I could even say that I enjoyed these foods more than foods I would choose on my own. The SGCPs were a context in which I frequently felt enjoyment over the experience of shared food (even as I was aware that others might not all equally or similarly enjoy the experience as I did).

Still, there were times over the course of this SGCP research when I definitely would have preferred to not eat the food presented to me. Several of these times were in SGCP classrooms, involving food that students had prepared. While the vast majority of dishes I ate within the context of the SGCP classrooms ranged from good to incredibly delicious, there were a few dishes that I felt nauseous trying to eat. In one dish, students had not adequately washed the sand from some leeks that we put in a soup. While few of the students seemed to mind or notice, I could not finish my bowl, and had to dump it in the compost bin. At another point, several students were coughing and sneezing into their hands, and then cutting the vegetables for a salad. Though hygiene was stressed in the kitchen, and this type of incident was unusual, the cooking teacher happened to be absent from the room at this time. As a guest I did not feel comfortable telling students to wash their hands, and so I let this continue and did not look forward to the salad. In the end, I did eat it, but the *knowledge* of risking a possible cold or flu made the dish taste rather unpalatable to me (a truly socio-material moment).

Outside the context of social eating events, when I was at the store or market by myself, my food decisions varied according to what I could find in the place(s) where I was shopping, and what sort of financial capacities I had at that moment. While I consider myself somewhat of a "foodie," seeking out local and organic labels, I was also on a limited research budget, and thus cost frequently entered my mind. For reasons of cost and (lack of) culinary knowledge, I also rarely bought meat to cook, though I would eat meat that was offered to me. I more often ate vegetarian and vegan meals. I also was particularly conscious of eating a variety of grains, and enjoyed experimenting with the fermentation and sprouting of wheat, oats, and rye. I was conscious of food allergies, including especially gluten intolerance, which is part of what prompted the grain fermentation experiments. Overall, I found the SGCPs to produce food items that I would probably choose to make myself, with few tweaks; indeed I frequently took recipe ideas home with me (e.g. for pear salsa, vegetable stew, fruit turnovers, or homemade pizza).

Beyond these preferences, there are two particular food trends that I was and still am most adamant about resisting: the low-fat craze, through which North Americans learn to avoid butter, animal fats, and other saturated oils, and the low-calorie craze, through which artificial sweeteners like sucralose, aspartame, and saccharin end up in both adults' and children's food supply. I consider these trends to be critically damaging to the overall health and well-being of North Americans, both in what they imply about correct body size (i.e. slimness is best), and in what they do, or more accurately what they do not do, to nourish human bodies (i.e. they are defined not by food itself but by the absence of food). Certainly, I carried these personal judgments about such food trends into the SGCP classrooms. I was worried about how the SGCPs might reinforce (either actively or by omission) a low-fat and low-calorie food culture, and I was intensely critical of the SGCPs on these grounds when I found evidence of this culture's presence. Nevertheless, I also found it difficult to pinpoint where exactly these ideas were coming from. Notions of good foods and bad foods, or of correct and incorrect body size, certainly circulated within the relations and interactions of the SGCP participants (both teachers and students), but the question of who was instigating the entry of these ideas into SGCP classrooms is less clear; was it parents, television, teachers, peers, the school board, the federal government? In this sense, the reproduction of expert-based knowledge systems within the SGCP is less straightforward than my initial stance allowed, and with a lot more disruption than I had expected. This broader story of "food knowledge" within the SGCPs is the subject of the next chapter.

3   Knowing food

Take the blue pill and stay here in the fantasy. Take the red pill and I'll show you the truth.

(Moofius in *The Meatrix*, 2003)

As previous chapters have discussed, activists within the alternative food movement have often framed their initiatives as enlightenment endeavors, where a select few are "in the know" and thus need to educate the unknowing "others" about what is best to eat (Guthman 2008b). The widely acclaimed short film *The Meatrix* exemplifies this type of a enlightenment framing to an extreme: a happy-go-lucky pig is introduced to the real world of industrial farming by "Moofius" and his entourage after choosing between a blue pill, which would keep him in his fantasy land, and a red pill, which would show him the truth (Meatrix 2003). This type of framing has been quite effective at exposing some of the ecological and social damage that industrial methods of agricultural production and distribution have generated, including pollution of waterways, rampant antibiotic use, and disruption of small-scale agricultural livelihoods. Indeed, it has worked successfully to counter the hegemonic power of agribusiness by (at least) bringing these issues into public view. Nevertheless, in presenting alternative food knowledge as something that is obtained rather than produced, this type of enlightenment endeavor also threatens to eschew the diversity that is present within food-based knowledge(s), creating yet another type of food hegemony.

Furthering this enlightenment push within school-based food initiatives is the field of nutrition science, upon which the nutritional guidelines of many government programs and school food policies are based (Nova Scotia 2006, USDA 2008). Because nutrition science is also predicated on the notion that knowledge is discovered, rather than produced, school food policies are often driven by a set of nutritional "shoulds"—lists of foods that a person should or should not eat. Although in some schools the simple "good food /bad food" dichotomy has been downplayed (particularly because of concerns over schools perpetuating eating disorders and body image problems), in many cases the dichotomy has simply been replaced by a less pernicious sounding set of labels: "anytime foods" and "sometime foods," (Leader, NS), or "Maximum, moderate, and minimum nutrition" (Nova

Scotia 2006). These new labels still carry the assumption that experts can dictate "best" nutrition practices, and that what is good for one is good for all, regardless of gender, race, class, age or any other (intersectional) social difference.

In this chapter, I dig deeper into the questions of what/who constitutes "legitimate" food knowledge, and why and how the process of knowing food *matters* to the success of SGCPs. My intention is not to dismiss the claims of alternative food advocates or nutritionists, for I feel to do so would be a disservice to all of the people who struggle daily to feed themselves and their families in more nourishing ways. There are real and important ways in which the industrial food system does not adequately contribute to the health and well-being of North American families and communities, and it is important to make these ways known. Yet, I do want to suggest that SGCPs need to move beyond the goal of enlightenment in order to really effect meaningful social change—most significantly because an enlightenment goal ultimately frames food choice as a matter of *individual* responsibility. Indeed, as the final segment of *The Meatrix* offers: "It is you, the consumer, that has the real power . . . it is up to you" (Meatrix 2003). While the knowledge that alternative food activists and nutrition scientists offer is significant, it is at best a partial knowledge: one that is *not* universal but rather situated within particular social, geographic, and bodied contexts that enable and constrain how it is taken up and acted upon in a variety of ways (Hayes-Conroy and Hayes-Conroy 2013). Given this, it is important to understand how SGCPs frame and approach food knowledge, and how SGCPs enable and constrain its production.

The chapter is broken up into two sections: knowledge as limiting (or in visceral terms, "chilling"), and knowledge as enabling (and perhaps also "charging"). In the first section, I explore the ways in which the case study SGCPs limited students' and their families' access to alternative food by furthering an enlightenment-based model of food intervention that disengaged them from the processes of knowledge production. Here I explore three instances of limitation: the perpetuation of universal knowledge claims, the construction of school and home as antagonistic spaces of knowledge, and the furthering of disempowering motivational drives. In the second section, I then move on to explore the ways in which the SGCPs enabled students and their families to engage in the production of food-based knowledge by allowing flexibility and diversity in the classroom. Here I explore three instances of enablement: the disruption of hierarchy through opportunity, the encouragement of student agency and peer-based knowledge sharing, and the application of food knowledge as creative potential.

Knowledge as limiting

In preceding chapters, I explored how knowledge is a visceral matter: how bodies *know* alongside and in connection to the mind in ways that ultimately call into question the boundary between the two. Furthermore, I stressed that visceral knowledge is a matter of lived experience: something that is articulated through the varied developmental processes of the body, and something that in turn influences how we (differentially) experience future life events. In these ways, knowledge can be considered a contextual and embodied practice that both defines our daily experiences and drives our daily actions—including what and how and where we eat. Further, because the outcomes of such a practice are not pre-determined, knowledge is also continually questioned and revised during this process—including our established tastes and preferences. Following from this, the section below explores how enlightenment ways of knowing in SGCPs can limit students' access to alternative and healthy food experiences by ignoring what students already know, and by prohibiting their own understandings of food from informing the processes of knowledge production.

Expert and universal knowledge

By September of 2007, I had engaged in 2 months of gardening and cooking with students, teachers, and parents from Plainville School. I had lived and eaten with many food activists, and at the urging of a local farmer and activist, had taken up experimenting with the fermentation of dairy, grain, and vegetables. In the middle of September, carrying all these experiences with me, I went to visit two local dietitians who were engaged in trying to promote school gardens in Nova Scotian schools. As I entered their office building, I was struck by "how different the space felt" from the spaces in which I had been researching and living (Field notes, September 2007, NS). Their office was housed within a medical center, at the end of a long bright corridor that had "not even the slightest suggestion of food: no smells, no pots steaming, no sounds of plates clanking" (Field notes, September 2007, NS). It was a sterile environment, isolated from food as a socio-biological practice. In this space, food was treated like medicine.

It was during my interview with these two dietitians that I began to realize how different food is in the places where people make it, and eat it, from the places where people make claims about it. In claiming food, food becomes abstracted, universalized, and made the terrain of specialists or experts. What emerges from such spaces is a fixed set of information: policy recommendations for what to serve in schools, tips on how to get students to eat more healthily, and lists of what constitutes maximum, moderate, and minimum nutrition. In the alternative food world, a slightly different list has emerged, but it is a list nonetheless, abstracted from the spaces in which we actually make food decisions. We are told to eat locally, seasonally, sustainably, organically, naturally, consciously, and slowly. In this abstract space, "food [becomes] our common ground, a universal experience," (Beard 2007), even though in our lived realities it is everything but that.

What happens when SGCPs combine these two food models from the worlds of nutrition and alternative food? What is precluded from happening when food knowledge becomes a fixed set of ideas? Over the course of my fieldwork, leaders and teachers often articulated an enlightenment-based model of food knowledge, explaining that SGCPs "are really about reaching people, those people who don't understand . . . who don't have a love of eating" (Leader, NS). To such leaders, being knowledgeable means knowing and being attached to food *in a very specific way*. Having "a love of eating" in this case does not and cannot mean loving McDonalds or Cheetos, it means loving fresh fruits and vegetables from the garden. To love McDonalds is to not *really* love food, to not *really* understand. As another SGCP leader explained, fast food represents "a sort of trough style eating . . . eating to live, versus living to eat" (Leader, NS). Fast food in this view is too quick and too processed to constitute a thoughtful, rational choice. In making such claims, a category of people defined as food "experts" emerge—those who know the truth about food, and who have the wisdom to tell the rest of us what we should and should not eat. In SGCPs, the "experts" are those in a position of power in the programs: the leaders (from ideologues to administrators), and also perhaps the adults who teach and volunteer in the classrooms. Indeed, students often came to view me as an "expert" of sorts.

In many of my interviews with SGCP leaders and teachers, the distinction between good foods and bad foods was quite clearly drawn in both discursive and visceral terms. In these discussions, it was clear who was in the know and who was not:

> When we [started this program], it was fried chicken, McDonalds, you know, garbage.
>
> (Leader, CA)

> When I came here, the [school] food was the worst in the nation. Extreme burritos, chicken nuggets, corn dogs, pizza pockets, all frozen crap.
>
> (Leader, CA)

> It is all pretty set [in the curriculum] . . . we present the info [about eating healthily], and then we see how they can incorporate that into a recipe.
>
> (Leader, CA)

> There are three groupings, Maximum, Moderate, and Minimum. We want to work towards 100 percent whole grain, no hot dogs, and look at the fat and salt, no processed foods. We allow cheese because you don't want to limit fat too much with kids, but we don't have meat because really [most kids] have too much meat everyday anyway, so I don't think they need it at lunch.
>
> (Leader, NS)

> It was sickening, what we saw going into students bodies.
>
> (Teacher, NS)

The above quotations illustrate how "healthy," "alternative" food has become a fixed category in these SGCPs, with distinct and rigid boundaries that are policed through both the rules and language of experts. As food knowledge is fixed in this way, it also becomes naturalized as a category beyond social critique. Food becomes the same for everyone, everywhere, not because everyone currently eats in the same way, but because it is assumed and accepted that everyone really *should*. According to the leaders and teachers quoted above, hot dogs and pizza pockets are unquestionably bad for everyone, being equivalent to garbage or crap, whereas 100 percent whole grain is naturally a universal good.

Of course, SGCP leaders and teachers are not the only ones making these claims. In fact, most students who I talked to about food knowledge claimed multiple experts: parents, television, Internet websites, the back of cereal boxes, and their community at large, in addition to their cooking and gardening teachers. Nevertheless, the SGCPs became for the students an educational space in which nutrition assumptions are confirmed. When asked about what constitutes healthy food, most SGCP students that I talked to would (initially) offer very similar answers—answers that would suggest to SGCP leaders that the program had been successful, and that students were learning the "right" information. It is these learned *shoulds* that in turn defined for the SGCP leaders the "universality" of food—the commonness that binds different SGCP students together. Indeed, in my interviews with SGCP leaders, many imagined that it was the universal goodness of such healthy food that provides the opportunity for a "shared" experience. Food, when it is "right," becomes our common good:

> [People need] to understand that food is the one thing we have in common, besides sleeping.
>
> (Leader, CA)

> Putting people together at a table can overcome differences, if you focus on the commonalities, on sustenance, you can move beyond petty differences.
>
> (Leader, NS)

In these examples, the nourishment that comes from SGCP food is assumed to be the same for everyone at the table, allowing students to overcome "petty" differences and focus instead on the common "shoulds" that bind them all together. In this line of reasoning, food knowledge is not something that arises out of difference, but rather something that erases difference. Goodness is to be accepted as fact. Students are not expected to share what they bring to the table (or even to bring anything at all); on the contrary, they are required to put aside their personal experiences and tastes in light of a common good. Following this line of thought, it does not matter if you are black, Latina, female-identified, poor, young, or middle-aged, because healthy/alternative food is "good" for you regardless; nutritionists, experts, and even nature itself tells us so. Indeed, SGCP food is deemed as "good" ultimately because our taste buds will supposedly confirm that this is true—at least if they are working correctly. Thus, the success of the SGCPs is measured in terms of changes in tastes, or developments of the palate, that will bring us closer to such (natural) unity, rather than allowing us to explore our social diversity:

[The SGCP] has refined their palates, [which] were unsophisticated.

(Teacher, NS)

School lunches have improved dramatically, and the kids know the difference; they have improved their palates.

(Leader, CA)

The program has increased his palate. He will eat more vegetables now.

(Mother, CA)

In the above examples, development of "taste" is singular in its developmental path; palates are "improved" from a state of unsophistication and narrowness to one of refinement and acceptance. Because, as discussed in previous chapters, such changes in taste are perceived as ultimately natural and inevitable, the SGCP leaders often considered these shifts in taste to be unproblematic. They were not seen as a matter of homogenizing assimilation or cultural imperialism, but simply a matter of natural development. Furthermore, such shifts in taste were considered *easy* for all students to access, since all it requires is the use of one's tongue. In this imagining, issues of cost, geographic availability, cultural difference, or racial identity ultimately don't matter; all students have to do is "use their senses." In this way, "no one has a hands up on anyone" (Teacher, CA).

Yet, my empirical evidence suggests otherwise. Students came to their SGCP classrooms with plenty of visceral topography; their senses did not lead them all in the same direction, and their "petty" differences were not overcome by a shared meal. For many students, the goodness of food was not something that could be defined solely by their SGCP class, but was instead something that is experienced as negotiable and differentiable among families and communities. A good example of this comes from SGCP students' reactions to the vegetarianism of their cooking classrooms. In both Nova Scotia and Berkeley, the absence of meat from the kitchen was something that some students lamented, while others celebrated. In the following conversation, four students in Central School were discussing whether or not it was appropriate to have a meat-free SGCP classroom. I asked them to reflect upon what it was like to not have meat in class:

Well, I am a vegetarian; eating veggies makes my body happy. I am happy [without meat].

(Student, CA)

But to some people meat is like their important thing to eat in life, like other than, I mean, to go along with fruits and vegetables.

(Student, CA)

Yeah, for some people some families have meat traditions and so those people get excited about that, and cutting meat out can get them less excited.

(Student, CA)

I try to eat as little meat as possible because I think . . . it's just in America where people eat a lot of meat. This video we watched in class, it said people in Asia don't eat as much meat, they have a lot of fish maybe, and they [don't even] eat . . . that much fish, like sushi.

(Student, CA)

The lack of meat in the SGCPs was a frequent complaint voiced by many of the students who ate meat at home, an issue that I discuss in more depth below. But the point here is that far from being equalizing, a vegetarian classroom exposes and highlights students' varying levels of experience/knowledge with regard to SGCP food. Some students' preferences are confirmed and legitimated by school policy, while others are refuted. When students and parents voiced complaints over the type of food cooked and served by the SGCPs, however, the typical recourse of SGCP teachers and leaders was to call upon the "experts," via government guidelines, scientific findings, or public health dictates, to make the rules seem universal. In other words, it is not the SGCPs that are making the rules; these programs are simply messengers of the truth:

You take away the personal nature by [saying] there are guidelines to follow, this new food policy . . . we can't serve [unhealthy foods] anymore.

(Teacher, CA)

Some students and parents don't understand why we can't have hotdogs for [even] one day, for field day, but they are coming around. The publicity on obesity helps, I think.

(Leader, NS)

We like pepperoni pizza, but we can't have it because it isn't healthy. [The teacher] told us that it was the school board that made the decision.

(Student, NS)

In none of the gardening and cooking classrooms in which I participated was there much to any direct nutrition education, although nutrition was frequently discussed in an informal manner between peers and teachers alike. Both SGCPs were also linked with more formal nutrition education curricula in students' homeroom classrooms, in which students learned everything from the function of digestive organs to the four major food groups. Because teachers and leaders often discussed and framed SGCPs as instruments of nutritional truth (via school policy, or nutrition guidelines), this educational setting helped to further the assumption among students that there are fixed "right" and "wrong" ways to eat. Within and beyond SGCP classrooms, students engaged in self-policing frequently. In my interviews and participant observation, students continually evaluated food along such fixed lines, and sometimes in ways that were more stringent than the SGCPs themselves:

Field notes, September 2007, NS

We made pizza today. I was in the kitchen with Ms. Dora and three new students, Beki, Fred, and Kayla. We made the pizza dough from scratch, and then the sauce too, blending in the vegetables again to "hide" them from the students. Beki grated cheese (a lot of it) to top the pizzas with . . . When they came out of the oven, they were really melted and cheesy. Fred and I put slices on plates for the Grade 1 students who had ordered, and we walked over to the classroom to deliver it. He looked at me and said, "I don't understand, I mean, they want us to eat healthy, but look at all the fat." He pointed to the grease on the pizza and made a disgusted face, shaking his head.

The SGCP students learned to evaluate food in terms of individual opportunities to make the "right" decisions. Eating high-fat foods, or foods with grease or butter, were considered bad choices, while vegetables are good. Food became "common" as children learned these rules and incorporated them into their bodily routines of perception, e.g. seeing pools of grease equates to feelings of disgust—a perception that I witnessed quite often in both SGCPs. It is in coming to know food in these particular ways that students were then able to bond over shared experiences and beliefs. Still, not all of the SGCP students were eager to take on such ways of knowing. Indeed, some of the SGCP students also often resisted these guidelines, insisting that such conceptions of good and bad were too narrow or constricted:

> There are times when I am like, this needs more oil and salt, and it is not going to hurt me to put it in! I ask if I can add more, but they are like no, we want it to be healthy.
>
> (Student, CA)

> I think it would be better if we were able to use butter, but not all the time. So it is not, I mean, we don't have to be healthy all the time, because a little bit of butter is not going to kill you, especially if you eat a salad.
>
> (Student, CA)

> How are we going to learn to eat healthy if we never have a choice? . . . I mean, if the only options are like healthy choices, how will that teach us how to make good decisions? I think we need to have more variety of junk food and good food, because that's what the real world is like.
>
> (Student, NS)

All of these students articulate a struggle with SGCP food guidelines. While these students still discuss food in terms of good and bad—a healthy "salad" versus a "bit of butter," they also suggest that they know food in ways that go beyond these guidelines—"it's not going to hurt me" or "we need variety." In this sense, these students are struggling to find a place at the SGCP table that will allow them to be more than what they sense the SGCPs want them to be. Further, as the last excerpt by a student from the Plainville SGCP suggests, such students are also trying to come to terms with their own agency in regard to food decisions beyond the SGCP table; they are struggling with how they can find their role in a world that is much broader and more uneven than the SGCPs themselves. Unfortunately, though, instead of using such struggles as opportunities to discuss difference in SGCP classrooms, or to complicate choice in the broader world, some of the SGCPs' leaders seemed to thwart students' struggle for agency by instead advocating a strong ideological push in the other direction. Rather than meeting students where they are, for example, one prominent SGCP leader insisted:

> I won't go into the cafeteria of that school because it is contaminated in a certain way with fast food values, I don't want to go there. I want to bring the kids into a park and sit at a table and set the table differently, so it is a surprise . . . They know when people care about them and don't care about them. You get it by what you put at the table.
>
> (SGCP leader, CA)

At this leader's table in the park, kids do not have agency; they are offered food of her choosing, in a place where she decides. Not only do comments like this disempower students from playing an active role in the production of food knowledge, they also deny the diverse realities that the SGCP students face in regard to food choice/practice. For example, while the ideologies of SGCPs often insist that a table is the most appropriate place to eat, in many students' lives beyond school a table is just one of many potential places to have a meal. According to the above comment, though, only the meals occurring at a table would be acceptable by SGCP standards. The others are "contaminated" by a different set of food values that are antithetical to her own, and thus in need of change. By policing the boundaries of good and bad food practices, such SGCP leaders end up making not only scientific and political assessments of students' food choices, but ultimately moral judgments about the students' broader familial and social lives. Thus, to sit at a table with one's family is to be cared for. It is, according to leaders, a natural situation, because it is something that the students will simply *intuitively* recognize as right and good. At odds with this notion of care are those who, by contrast, seemingly "don't care," the "others" who are responsible for feeding children: parents, grandparents, guardians, and friends, but most often in my SGCP research, young mothers. This conflict between school and home is the subject of the section below.

The knowledge of school and home

Largely by way of the enlightenment frame of SGCPs, the spaces of school and home are sometimes constructed as antagonistic. If food practices at home do not conform to those in SGCP classrooms, the two spaces are often deemed to be in conflict. This was true in my interviews with adults on both "sides" of the divide: parents and guardians on the one side, and teachers/leaders on the other. Nevertheless, these "sides" are not equally constructed. SGCPs have the backing of government policy, nutrition science guidelines, and alternative food advocates' claims to naturalness. Further, those involved with SGCPs are said to "care." In contrast, parents who deviate from the accepted best practices of SGCPs have none of these supports; and it is they who must defend their "uncaring" actions.

Leaders and teachers of the case study SGCPs frequently discussed parents in terms of an obstacle or difficulty. When I asked them about the problems that SGCPs faced in the future, the role of parents often came up. "Really, parents have been our biggest hurdle," said one SGCP teacher from Plainville School, "but they are coming around" (Teacher, NS). Another leader explained: "The idea was to influence the parents by getting to the kids. There is more bang for your buck that way, because you are reaching them early in life" (Leader, NS).

Others took a more dismissive approach to parents, suggesting that there was a futility in attempting to bring them on board. For the SGCP teacher below, the goal was simply to have an influence on student decisions in the future, after they leave their homes: "That's all you are hoping for, you plant a seed, and when they are an adult [on their own] they will remember" (Teacher, CA).

During my research, a frequently repeated sound bite among SGCP leaders and teachers was "get to the kids early" (Leader, CA). The reasoning behind this was multifaceted; early on, students have less developed tastes and preferences, and introducing kids to vegetables when they are young means a better rate of success in later years. If students became used to growing and consuming fresh fruits and vegetables in elementary school, by the time middle or high school rolls around, the whole process would be second nature. While such a chain of events sounds plausible, the implied assessment behind this pedagogical sound bite is that the parents are currently not doing an adequate job at home (whether simply for lack of care, as was often implied, or for more dire reasons of financial and time constraints). After all, if parents were doing what SGCPs do then getting the kids early would not be such an issue. I brought up this dilemma in an interview with a food activist in Berkeley, and he confirmed my interpretation, joking: "Getting the kids early . . . yeah, everyone says that. Kindergarten isn't early enough. Take them out of the home when they are born [laughs]" (Activist, CA).

Of course, to many parents this is no laughing matter. Some of the mothers who I interviewed did not find it at all easy to feed their children. Some struggled with financial or time constraints, others with "picky eaters" or food allergies. Most mothers (and some fathers, though I spoke mostly with mothers) described a home practice of food that in some way or another differed from school in minor or major ways. From ethnic traditions to veganism to "meat-and-potatoes" households to takeout pizza, a lot of families did not practice the same food habits as the SGCPs encouraged. The judgment that several of my interviewees leveled against parents for this lack of congruence was harsh and unforgiving:

[The parents] eat carrots and peas and that is it. Broccoli? [they ask,] what is that? No thanks [they won't eat it] . . . It is a lack of intelligence or education.

(Leader, NS)

The kids. . .are hungry for food, but they are hungry for someone to care about them, so this food comes with care and a whole different set of values. With authenticity.

(Leader, CA)

In these two passages, leaders blame the students' primary care givers (most often mothers) for not doing an adequate job of caring and being a role model for their children. They question parents' "intelligence," or the sincerity of their ethical practice, and imply that SGCPs can do a better job at parenting, at least around issues of food. SGCP leaders, however, were not the only ones to level such criticisms. I heard similar assessments from parents and students:

JHC: Where is the problem coming from?
Student: Some kids just eat crap for breakfast.
JHC: Why?
Student: Because their parents don't care and they aren't old enough to be exposed to good food.

(Student, NS)

I noticed on some field trips there are kids with lunches that I am like, fire hot Cheetos and a Pepsi, and then this kid from Pakistan had his mom's home-made Pakistani food. So some were really good and others were like, oh my god, mom is poisoning you!

(Mother, CA)

Hopefully what the program is doing is changing cultures. I mean, it is definitely changing families, but hopefully cultures. [My] kids have grown up with the *truth*, real good information. There are other families that this isn't true for, and you can see it, they are all overweight, they are huge, and they have health problems. But my kids look very healthy.

(Mother, CA)

In these assessments, we can see that judgments leveled against certain parents do not just arise from the leaders of the SGCPs, but from student participants and other "on board" parents. This trend reflects the broader social assumption that food choices are ultimately a matter of individual or familial responsibility. In this sense, we could say that the SGCPs are not doing enough to counter this assumption, or to teach about food choices within their broader social context. For example, I did not hear similar judgments voiced against the industrial food industry, the neoliberal economic structure, or the socio-spatial inequities of health care provisioning.

In many of the above statements, there is an underlying implication that SGCPs are therefore *particularly* important because they can help students whose parents' food practices and tastes differ in "bad" ways from SGCPs. In this view, by introducing such students to new kinds of food/knowledge, SGCPs can perhaps undo the bad habits that such parents have imparted to their children. Nevertheless, my interviews with parents, teachers and activists suggested that it is not always these youth that are being "helped" the most by SGCPs. In fact, several interviewees suggested that the SGCPs helped instead to further bolster the kids whose parents were already "on board":

> Families who have more food culture around fresh ingredients are probably benefiting from the program more because the kids know [those ingredients] and they can make the recipes we make at home.
>
> (Teacher, CA)

> My kids get more reinforcement at home than other kids … They were watching something on television about the health risks of meat and they said to me, "But mom, why don't they just use TVP [textured vegetable protein]?" So, the program really solidifies what I teach them at home.
>
> (Mother, CA)

In both of these examples, it is the children who already eat similarly to the SGCPs that are benefiting the most, by having their home food experiences confirmed and legitimized by their school's practices. Many of my discussions with students also confirmed this story, suggesting that what students learn in the SGCPs is not equally transferable to every student's home life. These discussions frequently revolved around differences in ingredients, and particularly meat. Students suggested that if they learned how to cook with meat in their SGCP classes, they would be more able to transfer what they are learning to their home practices:

> I mostly eat meat in my house. My family is all about meat. If they taught us nutritious things about meat in the cooking class, we could use [that information] at our house.
>
> (Student, CA)

> The food we cook at home isn't as healthy, I am gonna say. They don't let us have meat here, and like if we was to make kale, we'd put some meat in it. And theirs is organically grown, which [we don't have] … It feels weird.
>
> (Student, CA)

> We eat deer at home, but we can't cook any meat here, because there is a policy or something, I am not sure why. I think it is not supposed to be processed, like by machines and stuff. Sometimes I think it is nice to have an equal amount of meat and vegetables. There should be some vegetables [in the SGCP], but not too many.
>
> (Student, NS)

By disallowing meat and other "bad" ingredients from the SGCP kitchen classrooms, the cooking lessons offered in these SGCP spaces often end up being more relevant to the students who already eat similarly to what the programs teach. The students quoted above are less able to use the recipes and lessons at home because they often cannot find the same ingredients in their homes, and because their schools do not teach them how to work with what they have— e.g. how to incorporate meat into the vegetarian dishes that they cook. Beyond issues of physical availability, however, there are broader issues of access here. This food is not just unavailable to some students because their parents cannot or do not buy it, but also because of the tension and confusion that emerges when school and home food practices conflict. Both students and parents from such families articulated feelings of discomfort, anger, frustration, and hostility towards their SGCP initiatives, particularly in (re)action to the perceived judgments coming from SGCP leaders:

> Food is a really personal thing for a lot of parents, and they take the [school food changes] as a personal attack, like they are being judged, or they aren't good enough parents because they give their kids pop.
>
> (Mother, NS)

> Some people get their bird up, like you aren't gonna tell me what to put in my kids lunch. I mean, I don't think it hurts to have a hot dog once a year, but I understand the school is trying to be a good model, so I dunno.
>
> (Mother, NS)

> One thing we haven't addressed in schools is the parents. Because there is this whole dynamic when the kid is being taught about [fresh foods] at school and then comes home and the single mom pops a frozen pizza in the oven, and the kid says something about how this isn't what she is supposed to be doing. Of course, the mom is just kind of really not turned on by the whole thing.
>
> (Teacher, CA)

In these examples we can see that there is a broader visceral barrier to taking SGCP ideas home, or to getting "on board" with healthy alternative food, that has to do with feeling judged or being told what to do. Because the SGCPs are framed as messengers of the truth, there is no space for flexibility and negotiation between school and home. In this situation, not only do students lack agency in the production of food knowledge, but parents lack agency as well. Notably, both of the above quotations from mothers came from participants who had a somewhat favorable attitude towards the SGCPs, and who were discussing other parents' (re)actions. The parents who *really* did not like the program also did not want to talk to me about it, most likely because I too was perceived as another food "expert"—someone who would place further judgment on them.

Unproductive motivation

A third way that the production of knowledge within SGCPs can be limiting to students is through what I call unproductive motivation. Although motivation tends to be considered an energizing and positive visceral drive, motivation can also be unproductive when it reproduces social norms that are damaging and restrictive to students' development. While I make this claim with the knowledge that the definition of "unproductive" is itself necessarily hybrid, situated, and negotiated in particular contexts, I think it is helpful to be explicit about what I want and hope SGCPs in general to be working towards, and thus what I mean by progressive social change. In the case of unproductive motivation, then, I base my understanding of what is unproductive upon what I, along with many other scholars and social theorists, have expressed concern over, e.g. trends that signal the reproduction of neoliberal subjectivities, the normalization of inequitable or oppressive gender roles, or the reinvention of social inequity on the basis of class or racial identity or body size. Along these lines, there are several types of unproductive motivation that I particularly witnessed within the SGCPs, including desire for capitalist gain, pressures of conformity, and competitive one-upmanship. Here I want to talk briefly about one type of unproductive motivation that I found especially troubling: fears related to body image and fatness. I also return to this topic in more depth in my discussion of gender and dieting in Chapter 6.

Fear of fat was the most obvious of the unproductive motivational drives that I witnessed during my research, and was the most frequently mentioned within the SGCPs classrooms. As I already discussed in the sections above, SGCP students' evaluative standards of what constituted good and bad food typically revolved around the presence or quantity of fat in their food. Vegetables and fruits were deemed healthy because they were low in fat and were fresh. Cheese could only be healthy in small quantities, because too much was fattening. Indeed, the manner in which most students (and some of their teachers) categorized food typically illustrated a fear of eating too much fat. Of course, it was not the *eating* of fat that was itself considered the problem. This fear of fat in food was translated into body image— a fear of *looking* too fat, or *being* too fat, and thus also being (supposedly) "unhealthy." When I asked students why they had such a SGCP initiative in their school, students told me that the objective was to make sure they kept a slim appearance:

> I think it is to help us not get obese when we are older.
>
> (Student, NS)

> It's because kids are obese.
>
> (Student, NS)

> Too many people are fat. They want us to be healthy.
>
> (Student, CA)

> Because we eat food that's hecka bad. People be getting obese.
>
> (Student, CA)

Although I did not hear teachers in either SGCP place a *particular* emphasis on fatness during their daily classroom activities, the students were certainly aware that the reasoning behind the SGCPs was to prevent childhood obesity. This was indeed common knowledge. Perhaps this widespread understanding was due in part to the large media attention in recent years that has focused on issues of fatness and dieting in general. To be sure, students often talked about reality TV shows like "The Biggest Loser" and dieting fads like the "South Beach Diet." Thus, as students engaged in activities within the SGCP classroom, students often tended to associate their gardening and cooking activities with the prevention of fatness. Notably, students frequently discussed their food choices not just in terms of the quantity of fat in an item, but also in terms of how it would make them look. Students articulated the idea that a thin appearance was desirable, and that looking fat was something to avoid.

The focus on fatness or *being* fat was not just something that I witnessed among students themselves, however, but something that seemed more broadly a part of the SGCPs programs as a whole. For example, in Berkeley one SGCP teacher mentioned that the cafeteria staff members were all given gym memberships at reduced rates. "It's about health with the staff too, some of the staff . . . are [too big], we all need to be good role models" (Teacher, CA). In Nova Scotia, one of the central reasons behind engaging students in garden-based activities was to encourage physical exercise, which would keep students slim and healthy. As one leader told me during an informal discussion in the garden, "It is a shame that so few students like to garden these days. But they will play video games . . . It is laziness" (Field notes, August 2007, NS).

Examples like these point to the need for SGCPs to pay closer attention to what motivation does beyond changing eating habits. The topic of social identity and food is a large one, but here it is important to consider how fixed ways of knowing food can translate into fixed ways of being and becoming, which ultimately preclude SGCP initiatives from being open to social difference. To the extent that students are motivated to all be the same type of person, or have the same type of body, the knowledge that SGCPs offer can be seen as limiting. However, SGCPs can also offer much more than this. The remainder of this chapter discusses the ways in which knowledge production in SGCPs can be considered enabling.

Knowledge as enabling

Because knowledge is a visceral matter, a matter of both physical body and intellectual mind, the outcomes of knowledge are not fully predetermined. While there are important ways in which the knowledge produced within SGCPs can be limiting to certain students, there are also ways that SGCPs produce knowledge that is enabling and animating to these same students. In the sections below, I explore some of the ways that SGCPs programs enable students to engage in the production of food-based knowledge, thereby encouraging rather than eschewing diversity. In particular, I want to discuss three instances of enablement in regard to food knowledge: the disruption of hierarchy, student agency and peer knowledge, and information as creative potential.

Disruption of hierarchy

> I hate shopping for food in the neighborhoods I grew up in, but only because I've traveled enough to know the difference. If you never leave the hood, you have no way of knowing how bad the produce is.
>
> (Kweli 2007)

The above quotation is taken from the website of Talib Kweli, one of several recognized "socially conscious" contemporary rap artists. I begin with his words because they were important to my own development as I engaged in this SGCP research. Kweli's album *Eardrum* came out in October of 2007, as I was transitioning from Nova Scotia to my Berkeley case study. The lyrics in this album, which are summarized in part by the passage above, served as my reminder that SGCPs are not just about limitations and restrictions, they are also about providing opportunity. Through listening to Kweli's words, I came to recognize that SGCPs could potentially help to disrupt the hierarchies that currently structure students' uneven access to fresh fruits and vegetables—in economic, geographic, cultural, and ultimately visceral terms. SGCPs can and do provide disenfranchised students with opportunities to relate to food and food practices that they otherwise would not have—thereby expanding students' relational reach.

In this understanding, it is not so much the food or practice that is the point, but rather the opportunity to relate that is itself significant. To focus on a particular food (broccoli) or food practice (sitting at a table) is to focus on the outcome— "improved" palates or habits. But the *opportunity* is something different. In an actor network sense, the opportunity to relate does not oblige a certain outcome of relating, but rather suggests an openness of possibility to what might be produced in the process of forming a relationship. In providing opportunities to relate, new types of knowledge can emerge. As Kweli's words suggest, this knowledge can be both devastating (in what it reveals) and empowering (in what it allows). A food activist in Berkeley explained her work in this way:

> The African community in South Berkeley has no grocery stores, not one in a 3-mile radius. There are 11 liquor stores, and fast food, Jack in the Box, and McDonalds. How can you read and learn and be a productive citizen if you don't have access to real food? We don't need scientific research on that; it is just basic common sense. So we said *we are going to provide opportunity and outlets*, not just bringing food to people, but helping grow that value of basic raw food, because it wasn't there.
>
> (Activist, CA)

When I began to look at the SGCPs in terms of what opportunities they provide, I discovered a whole new way to analyze the programs—even down to the visceral level of sensory perception or taste. I asked myself, what opportunities do the SGCPs provide not only to taste, but also to feel out new ways of tasting? In other words, if some students never have opportunities to taste certain foods, what might the act of tasting allow that would otherwise not exist? Of course, the answer to this question is not fully attainable because, as previous chapters discussed, the outcomes of tasting are not predetermined. Nevertheless, by discussing SGCPs in terms of opportunities, we can begin to imagine and to allow for the possibility of disruption and difference. In thinking about taste in these terms, comments like the following can carry new meaning:

> If you grow up in a family eating well, you may go to junk as a rebellion, but you will come back because it is engrained in you. But the schools can influence those kids who never got the chance to taste quality food at home; it can be an epiphany.
>
> (Teacher, NS)

> I don't think you can like something unless there is a sensory relationship to it. If all you ever eat is canned beans, why would you touch a [fresh] green bean?
>
> (Teacher, CA)

Rather than interpreting these comments in terms of their implied end goal (eating healthier food), I began to understand such claims as descriptive of the potentialities that SGCPs allow in bringing things into relation: bodies, foods, ideas, tables, etc. From this perspective, SGCPs are not about trying to convince "others" to eat in one particular way, but rather about the opening of possibilities for new ways of eating that could disrupt current uneven tendencies in our food systems:

> The [Center for Disease Control] says one in three Caucasian kids, and two in three African American and Hispanic kids will have diabetes in their lifetime. So, we do see the obesity crisis growing, and faster in communities of color.
>
> (Activist, CA)

> If I ask minority students what is the one thing we do here that you don't do at home, but you'd like to, the number one thing they say is eat together. It isn't even so much about food as the community and feeling a part of something.
>
> (Teacher, CA)

> I am a privileged, well educated and traveled man; I have seen what local food can do in Italy, but a poor underprivileged kid from rural Nova Scotia, where does he or she get that opportunity? So, that is the educational aspect for me.
>
> (Leader, NS)

While the SGCPs that I studied did not generally go beyond school to address the issues of economic or geographic availability that lead to such statistics, in bringing poor and minority students into relation with new foods and new food ideas, they also widened the scope of possibilities that were available to these students. "Possibilities" in this sense refers broadly to students' abilities to develop new visceral imaginaries—to have novel experiences with food/practices that allow them to disrupt current habits of bodily (re)action and begin to feel out different ways of being and becoming (and perhaps to "access" healthy food in their own way). This is different from the assimilationist interpretation of taste education because the focus is on the *disruption* of hierarchical patterns, rather than on the construction of a hierarchical end goal. For example, in many of my interviews, leaders, activists and students discussed the need to disrupt patterns of relating to fast and processed food:

> We have to make a huge effort to get out of the addiction, especially when they are trying to figure out every conceivable way to get you hooked. You got the ads on the bottom of your gym shoes . . . so we have to create events that reach people in all the seductive ways of culture . . . [because otherwise] you are at the mercy of fast food nation.
>
> (Leader, CA)

> Childhood memories are the strongest, and [that's why] you have to start early. Right now they are building their nostalgic connections with Cheetos.
>
> (Activist, CA)

> Like McDonalds is addictive, it is like smoking a cigarette, and you can't stop. Serious. It is like, McDonalds, they've got something in their food that's addictive.
>
> (Student, CA)

In focusing on the disruption of attachments or addictions, we can begin to understand how the idea of sensory education could be important to dismantling inequitable power structures; in these examples, this is true particularly in regard to our capacity to undermine corporate interests, advertising, and agribusiness. In this sense, these examples are not unlike what Thrift (2005) suggests when he warns of the capacity of those in power to produce affects that control our life practices or limit what we deem possible. With the above passages in mind, then, I want to take seriously the possibility that SGCPs could help disenfranchised students to "take back" their taste buds (Hayes-Conroy and Hayes-Conroy 2008)— not in any pre-defined way, but in a way that allows them to interrupt the economic and social forces that tend to keep them in connection with foods that they describe as addictive.

Agency and peer knowledge

> It is coming from her, she owns it.
>
> (Mother, NS)

> Kids often want to rebel against their parents at this age, so this way my kids can have it be their thing.
>
> (Mother, CA)

> It's our garden, we work in it, we go to school here, we plant the stuff, and then we get to eat it. It is ours.
>
> (Student, NS)

A common claim about SGCPs is that they "work" because they allow students to make healthy eating "their own." But, what exactly does this mean, and what processes and opportunities does this entail? Over the course of my research, I witnessed many instances where students' agency was encouraged in SGCP classrooms, and where students were invited and even expected to contribute their own ideas to a collective production of knowledge. One way that SGCP teachers encouraged student involvement is through the flexibility of their pedagogical style. Although flexibility was more easily accomplished in the kitchen classrooms, because the gardens required a more rigid and longer-term plan, both the kitchens and gardens of SGCPs were places where teachers embraced a flexible approach:

> There is a lot of flexibility in this type of [garden] work, letting them use their own skills and passions, and teaching to specific children, letting them shine where they can.
>
> (Teacher, CA)

> I am really flexible, that is my style. The kids take it seriously, but they also have a lot of freedom. We talk about anything in [the kitchen].
>
> (Teacher, NS)

In addition to being flexible, having freedom in the kitchen or garden classes is also a signal to students that they are trusted. The students that I interacted with in the SGCP classrooms were often most animated when they were asked to do something out of the ordinary—something that illustrated that their teachers had confidence that they could handle the task. For example, one day at Central School three students and I walked five blocks down the road to collect coffee grounds for compost for the school. We brought along a wheelbarrow, and came back with a good amount of compost material. The students were "visibly animated" (smiling, walking faster than usual) at the opportunity to engage in such a task, particularly because it required "going off school grounds in the middle of the school day"— something that was not usually allowed (Field notes, November 2007, CA). Teachers also discussed the issue of trust during interviews:

A lot of people who visit say, how can you trust the kids with these knives? But you can. The kids know how serious it is . . . and they appreciate us having enough respect and trust for them to use these tools. They live up to these expectations. We never have a problem, they don't mess around.

(Teacher, CA)

Trust was also something that came up frequently in regard to taste. Students seemed to enjoy the opportunity to alter recipes, to sample and make changes while cooking, or to have some say over the way food tasted. SGCP teachers often asked students to try a dish and figure out what they thought it needed more of— spices, salt, a certain ingredient, etc. When students are asked or expected to play such a role in decision-making, they are shown that their visceral judgments matter. Rather than legitimizing those visceral judgments that fall in line with certain accepted food habits or behaviors, in trusting students to make decisions about taste, students are shown that all experiences and judgments have value. In other words, it shows them that they can be co-producers of food knowledge: "We found that when we conducted taste tests, there was buy in. The kids felt their opinions were valued" (Teacher, NS).

In this way, the taste education that these SGCPs offer does not point towards some final end goal where every student will come to like the same foods. Instead, the pedagogy is one of encouraging different students to provide input and suggestions in producing the final meal or snack. The end result is not fixed but negotiated. One teacher at Central School discussed with me how this pedagogical strategy was a change for her. Initially, she had considered the work of SGCPs to be more focused on a fixed idea of health. However, after working in the school for a while, she came to the realization that fixity was not the best strategy for reaching students. She needed to recognize where the kids were coming from, and allow that to inform her educational practice:

When I first started I had a different vision, also for my own health I had a different vision, and I really believed that if you could encourage kids to eat more fruits and veggies and beans [that was the ultimate goal.] But as the years have gone on I find that I am trying to meet the kids closer to where they are coming from, so we may use butter or eggs or white flour, or some sugar, but good quality . . . I find with the middle schoolers a lot of lecturing about health doesn't impress them. It's not my style either.

(Teacher, CA)

Often in the SGCPs I studied, teachers not only encouraged but expected such active participation from students. Through such expectations, the SGCPs emerged as spaces where a diversity of tastes and experiences can and do exist, and where a diversity of students can and do take an active role in the production of food knowledge. In this sense, the "goodness" of SGCP food can be assessed by examining the *process* in which food/decisions are made, rather than by focusing on the final product. Several SGCP teachers focused on the idea of process over product during their interviews, explaining that SGCPs were consciously designed to bring together a diverse group of students, with varying ideas and experiences with food. The expectation was that through the process of such interactions, students would teach *each other* about food not in any pre-established ways, but rather in the diverse and haphazard contexts of their lives:

> There are different reasons why we have selected students to be in there. Some kids need to have training about how to cook for their own survival; others have a lot of background [with vegetarian food], others need a creative outlet. We want a heterogeneous class so that kids bring in different perspectives.
>
> (Teacher, CA)

The notion that SGCPs provide a space for students to teach *each other* is an important one. During my research I witnessed the SGCPs providing daily opportunities for peers to collectively engage in the production of knowledge, with little to no involvement by teachers or adults. The significance of this unsupervised knowledge production is two-fold. One, students are able to play an active role in their own education, which has significant impacts on their experience of the programs overall. Second, as a food activist in Berkeley pointed out to me, students of middle school age are often more likely to listen to each other than they are to an authority figure. As he explained of his experience working with kids, if your best friend wants to talk about healthy alternatives, you are less likely to feel judged or attacked by such comments, and are often more likely be open to hearing what s/he has to say: "When it is your peer, it is more of a conversation, it is more about sharing and less hierarchical. You are being taught, but there is a dialogue, a certain vernacular. It is easier to receive when you are dropping knowledge on a comrade" (Activist, CA).

In my observations of peer knowledge-sharing within SGCP classrooms, I witnessed a variety of different scenarios. At times students were gentle and forgiving with their information, offering suggestions and advice rather than strict critique: "Healthy eating isn't like never eating pop, it is about finding balance. I mean, I never really liked pop anyway, but if I did, that would be ok too" (Student, NS).

At other times students were harsh and abrasive to each other with their information: "Bull shit! Do you know how nasty [KFC] is? They breed their own chickens and they are born beakless and featherless. I read about it" (Student, CA).

Although peer relationships were certainly not free from hierarchy, they were undoubtedly strong forces within the SGCPs that influenced students' experience of food in powerful ways. The strength of peer relationships calls into question the centrality of SGCP teachers and leaders to positions of power within the programs. Often in my daily experiences with SGCP classrooms, peer relationships appeared more persuasive than anything a teacher suggested or demanded. As a teacher at Central School said to me, "If you can get a popular kid to say, 'This is hella good,' it is a powerful force at the middle school level." Indeed, it was for this reason that some SGCP teachers made rules about students saying negative or nasty things about SGCP food: "One of our teachers wants us to only say 'Hmmm, interesting' after tasting our food. She doesn't want us to use words like nasty . . . because you could be a bad influence on other kids. But that really bugged me, and none of us liked her" (Student, CA).

Despite some teachers' attempts to limit negative comments or interactions, however, they were a common occurrence in SGCP classrooms, alongside many other responses: positive, intrigued, aloof, and so on. Students frequently changed their minds in unison, and often quite dramatically, about what they would or wouldn't eat, or about what tasted good to them. This happened so frequently that I came to label this phenomenon "visceral peer pressure":

Field notes, September 2007, NS

All the kids were out in the garden, passing around a basil leaf and smelling it, at the request of their teacher. One boy made a face and said "Ewwww, gross!" as he passed it. After that, every student had the same reaction down the line, until one girl smelled it and said, "Oh! This smells great, what are you talking about?" So then the door was open again for a variety of reactions, and others agreed with her.

Whether or not visceral peer pressure is an accurate label here, what is clear from these accounts is that students influence each other a great deal within SGCP classrooms. In this sense, it is important to remember that a lot more occurs within SGCPs than nutrition lessons and gardening/cooking instruction. Whether direct student involvement is a matter of pedagogical strategy, a happenstance, or a force that is avoided, students play a large role in how food knowledge is produced and reproduced within SGCPs. Paying attention to this function is therefore important in understanding how SGCPs function, and what they are actually capable of doing in terms of their capacity to motivate.

Information as creative potential

The third instance of enablement within SGCPs is the role of information in allowing for creative possibility. This refers specifically to the fact that SGCPs provide students with information that allows them to (re)act and respond to their particular social contexts in creative and important ways. For example, over the course of my research, poor and minority students often expressed gratitude for what their SGCPs allowed as far as new ways of relating to and creating food:

It teaches me that I can do things in my own house, for myself, I can make food and I don't have to rely on fast food or on my family members.

(Student, NS)

People think that if you eat healthy it's nasty and there is nothing good about it, but here they show you how to put more flavor in your food, but it is still healthy.

(Student, CA)

Here again we can see that the production of knowledge in SGCPs is not only or always limiting to students. Knowledge can be enabling to the extent that it allows students to develop new visceral imaginaries that open up creative possibilities in the food spaces of both school and home. As the students above describe, their SGCP offers them the opportunity to relate to "healthy" food in new ways. The SGCPs show these students that they have different options in terms of both what they do with food (buying fast food versus cooking) and also how they feel about it (assuming that healthy food is tasteless versus expecting even healthy food to taste good). In this sense, students are an active part of the production of knowledge within the SGCPs; knowing food is not about holding the "right" information but about realizing how information can expand one's opportunities to know and create.

Students also discussed this creative potential in terms of the idea of flexibility, discussed above, and particularly in terms of what they were and were not allowed to do, or how much freedom they had in the daily operations of the kitchen or garden. Students often associated flexibility with a potential for more creative moments, where they could not only take control of the activity, but also use their own ideas and interests to steer the project in different ways. Students also found that this "freedom" was more available in the kitchen than the garden, particularly because of the increased opportunity for creativity in the kitchen: "Our gardening program is okay . . . I like our cooking class better [because] we have more freedom. I know that sounds dangerous, but it is not. They talk about safety before we cook" (Student, CA).

Indeed, one of the most surprising revelations over the course of my SGCP research was the extent to which the cooking programs encouraged students to think and act creatively in regard to food and food practices. This was true particularly in the Central School SGCP, where approximately every month the students engaged in a group cooking contest. On these days, students were given no recipe or instruction, but instead just a pile of potential ingredients, out of which they could make whatever they desired. The results were imaginative and diverse. With eggs students made omelets, fried rice, bread, and meringues; with beets they made salads, roasted vegetables, garnishes, and food dyes. The following field note excerpt describes one of these contest days:

Field notes, December 2007, CA

At the end of the day today, the group at table three was by far the cleverest and most amazing I had seen. Their classroom teacher even commented on what an opportunity it was to be trusted to make a dish all on their own, what that allows in terms of their own motivation and excitement. The girls were so excited about their dishes, and asked me if I liked the garnishes that they made (a bit of rosemary on every plate). Another group made corn chips for their salsa, by making tortillas, cutting them up, and frying them. Another boy carved intricate figures out of the radishes and an apple.

As the discussion above suggests, this sort of creativity that was encouraged within the SGCPs seems to reach over into other areas of the students' lives, beyond the cooking room and the school. Through these events, I began to recognize that the information and technical skills that the SGCPs offered students—everything from learning how to turn on a stove or oven to learning how to peel a winter squash—was perhaps fixed in the curriculum, or in the nutrition guidelines, but certainly not in the lived spaces of the students daily lives. The SGCP information was put to use by students in a variety of different ways, both within and beyond the school classrooms. When viewed as partial knowledge instead of universal knowledge, the ideas and insights that SGCPs offer students can be seen as potentially enabling of their creativity, adaptability and survival.

Overall, this chapter has discussed ways in which SGCPs can both limit and enable different students' access to new ways of thinking about, preparing, and experiencing food. The first section explored the SGCPs critically, questioning what constitutes and who creates legitimate food knowledge, and asking how some students might be precluded from participating in the production of food knowledge within the SGCPs. The second section built from this critique to bring back in a more hopeful interpretation of the SGCPs, in which the opportunities, interactions, and information the programs offered could be seen as potentially enabling, and thus also viscerally animating. These perspectives together are important to understanding the complexities of the SGCPs in terms of both what they allow and preclude. The chapters that follow in Part 2 of this book pick up on these threads of analysis through empirically driven examinations of SGCPs across lines of class, race, gender, and age-based difference. Although the following chapters consider each of these four categories separately (for the purposes of clarity and organization), it is important to remember that these categories are often lived not as distinct but rather as intersectional (Valentine 2007). Whenever possible, and whenever my empirical evidence demands it, I hint at these intersections; nevertheless, it is important to keep in mind that each of these four categories, at least in terms of the visceral, exists as a moment or impulse within a much wider, intersecting and relational web of forces that influence the what, how, when, where and why of food decision-making.

Part 2
Tasting difference

4   A tale of two dinners

This chapter centers upon a detailed account of a two-dinner evening in rural Nova Scotia, experienced through the eyes of myself—a researcher and food enthusiast—and a friend of mine, who is neither an academic nor a local (culinarily or culturally speaking). Through our interactions with each other and many others over the course of a long, food-filled evening, the two of us collectively uncovered the cornerstone of a local friction—the clash of two distinct, class-based food cultures. The two dinners could not have been more dissimilar. The first was rushed, awkward, and unpretentious, a potluck of casseroles and previously frozen peas, squeezed in between a non-alcoholic toast by the Chief of Police and a cookie-laden plea for donations. The second we experienced slowly and organically, over several rounds of wine, and no dearth of self-congratulating, as our meal literally and carefully migrated from garden to dinner plate. Both dinners took place in the same town—each labeled as a local get together—and yet, as we recognized acutely by the end of the night, a chasm separated the two. This chapter investigates and analyzes this chasm in an effort to illustrate how food practice and social class intersect. It also explores what might be done, in terms of the practices of alternative food, to lessen these social and visceral gulfs.

Food difference and social class

While racial identity (the subject of Chapter 5) was an obvious influence on food experiences and practices within Central Schools SGCP, class-based food identities were more apparent in my Nova Scotia case study. Rural Nova Scotia is racially homogeneous but varied in terms of levels of income and education, and indeed in terms of access to food (Williams et al. 2007). Class identity, however, can be more difficult than race to name and discuss, particularly because many North Americans consider themselves to be "middle-class," despite a tremendous amount of discrepancy in purchasing power, education level, and cultural capital within this social grouping (Walkowitz 1999). In my research, the class identities of research participants were bound up in a variety of different social characteristics, including level of education, occupation, family background, travel and worldliness, and also one's relationship to urbanity. They were also, as I will show, bound up in particular food preferences and practices.

In terms of alternative food, I have previously discussed how the alternative food movement is often coded and critiqued for its "elitism" (Leitch 2003, Guthman 2008a), which broadly signals that alternative food tends to be a practice of the upper classes, as opposed to what we might consider the working class, the lower middle class, and/or any person or family for whom food security is an issue. As I have argued, such identity-based attachment to particular foods can have strong implications for who tends to be turned on or "animated" by alternative food practices—or who can access positive visceral (re)actions to such food—and who tends to be "chilled" or turned off, unable to access such positive sensations. In what follows, I discuss how class-based food identities are both reinforced and resisted through SGCPs in ways that have implications particularly for how parents (re)act to the programs, as well as how the students themselves (re)act. To do this I focus mostly on Nova Scotia examples, although these issues were somewhat apparent in Berkeley as well. I contend that the SGCPs I studied further a "gentrifying" trend of the alternative food movement in regard to "simple" and "local" eating, while at the same time providing opportunities for the re-claiming of healthy food as a rural, no-nonsense practice of non-elites. I begin with a lengthy excerpt from my field notes that exemplifies the class-based tensions that surround food within my Nova Scotia case study site. In this excerpt, which I title "A Tale of Two Dinners," I juxtapose two of my own food experiences in Plainville in order to describe the class-based divide within the community, and to situate myself within this split.

The gentrification of simple eating

Field notes, September 2007, NS

A TALE OF TWO DINNERS

1.   FIRE HALL POTLUCK

I wanted to attend this fire hall potluck that I had seen advertised in the local paper because it was right down the street from Plainville School. I brought my friend Paulo along with me because I didn't want to go to the potluck by myself. I felt like I would feel out of place and uncomfortable. I had an invitation to a dinner later at Kim's, but no direct invitation to the potluck, so I wasn't sure what to expect; probably rural people and lots of food I didn't really want. I told Paulo this. We approached the fire hall and it looked pretty packed with cars. The inside, however, had lots of empty seats. I took a deep breath, switched on my social demeanor, smiled and asked how much it cost. 6 dollars each, I was told, which I happily paid to a nice man who talked to me about the weather. We scouted out places to sit. Paulo was such a great sidekick—not shy, but instead very social, despite our out-of-placeness. He was used to being out of place, I suppose, having moved to

Canada from Brazil less than a year earlier. We got in line and got paper plates and started dishing out food from the choices. There were lots of potato salads, macaroni salads, cold cucumbers and tomatoes (the only fresh veggie on the table), and some pickled green beans. There were lots of baked dishes, tuna casserole and lasagna, and potatoes in butter, rice with beef, chili, and things like this. Most of the people there had white hair and were probably in their 70s or 80s. There was one woman firefighter ("Cool!" I thought), and all the rest men. The men served coffee and tea, and there were (only) women in the kitchen, bringing out more food. The tables were set with white paper on top, and metal ware, with napkins. There were also plates of cakes on the table in addition to baskets with rolls (this seemed strange to me, but also a fun addition, especially since many others seemed so appreciative of the abundance of these cakes). There was also milk for tea and coffee. I would guess from the set up that they were expecting an older crowd, but also maybe a mixed crowd—both grandparents and their grandchildren. The cakes were chocolate, carrot and blueberry, and there were cookies too, some homemade and some store bought. And then a whole table of pies (Kim's group later said, "Oohhh," knowingly, when I mentioned these pies—I wasn't sure if this was because the pies were tasty and abundant, or because pies are considered "bad" for you). The fire hall was right next to the school, and I saw one woman who recognized me from the garden in the summer, and another young girl who I knew from the summer rec program. She waved. It was nice to see some people who I actually knew. Paulo and I sat across from an older couple in their 70s. They were very nice and told us we could sit wherever we wanted. So, we sat across from them and I started a conversation about how I was new to the area. We talked about their grand kids, and how they wouldn't come to this potluck supper because they were picky eaters. "So, what do your grandchildren like to eat?" I asked. "Steak, French fries, lobster. They eat their veggies too, though," explained the grandmother, knowing that this was an important detail. "They are in middle school and high school." I nodded. We engaged in small talk about the weather, weekend plans, etc. and they talked to Paulo too, about Nova Scotia and Brazil. They wished him good luck and told him to enjoy his time in Nova Scotia, and they also welcomed him to the area. Later Paulo told me that he felt like the pair was very genuine and sincere. The old man to my right didn't say much, but he knew the other couple through a religious network. Lots of people here know each other through church, the couple explained. The little kid down the table from us was too picky an eater, the woman supposed, and so he just had a piece of cake for dinner. Nothing else. The man who sat across from us had two pieces of pie, and the woman said to him "You go right ahead," as if he needed her permission to enjoy the sweets. Later, Paulo and I both agreed that the food served at the potluck wasn't very "good" by our standards: it was packaged and processed and not very fresh. But, we admitted, despite not loving the food's taste, we still thoroughly enjoyed ourselves. In fact,

I was pleasantly surprised by the warm welcome and the informal, laid-back tone. We ate off paper plates, but had real mugs for tea, which I thought was a nice touch. Someone would have to clean all of them, but I was sure someone had already signed up for this job. After eating quickly, we lingered over our tea and cake. Paulo and I talked about all of this afterwards: the food was not the main point, we agreed, at least not in terms of the details that I was so used to focusing on—origin, freshness, nutritional quality. It was more of a social time for people to get together and talk about their lives, and be a part of the community. The specifics of what they ate were secondary to this, though the food was seemingly both enjoyable and familiar to most. The purpose was also, of course, to raise money for the local firehouse, hence the entry fee. Some of the attendees were poor (mentions of working-class jobs, notably older clothes and cars) and some were richer or had more money saved (mentions of a summer house, or the leisure of retirement), but no one that we encountered accentuated or flaunted such differences. Paulo and I left with a wave and a nod to the couple we had just met and shared a meal with. We both felt pleased by the experience.

2.   KIM'S DINNER

After leaving the fire hall potluck and briefly showing Paulo the school garden where I was working, we drove down the road to Kim's house for another dinner party. Kim had invited me over for dinner (partly, I think, since I had mentioned to her before that I'd be alone this weekend. Her invitation was a gesture of kindness and generosity). But, I brought Paulo along, an unexpected and uninvited second guest. Nevertheless, both of us were greeted jovially by the hosts and the other guests, as well as the big family dog. We first sat on comfortable lawn chairs beside their garden, talking for a while about wine and grapes, and about other food-related topics as well. Other folks who I had come to know (as fellow Nova Scotia foodies) were there as well, with their partners or friends. I felt very much in my element— a bunch of foodies, talking food. Paulo, however, being new to the foodie world, had very little to say, and I started to feel anxious about how he might not be having a good time. After a while we stood up because the dinner was ready (our second dinner for the evening). However, Paulo and I hung back a bit because we weren't actually sure that we were invited for the dinner part of the get together (I had mentioned to Kim that I'd be going to the fire hall potluck first, and she had told me to come by to her party afterward, anyway. So, I wasn't sure if she had counted us in as guests for what would be our second dinner). But, I quickly counted the placemats and discovered that there were indeed places made up for us. So, I sat, and Paulo followed my lead. The dinner was almost totally from Kim's garden, as I expected, with the exception of a few things: garbanzo beans, olive oil and feta cheese. The whole meal looked delicious to me (and much more what I am used to than the fire house offerings). I took a little of each of the fresh veggie dishes (broccoli, green salad and green beans). I passed on the bread and the chicken (though it also looked wonderful) because I had already eaten and didn't want to look piggish. Trying to justify my veggie selections, I offered a comment to the table about how there weren't many fresh vegetables at the potluck, and Kim's husband said, "Yeah, well, veggies would be an anomaly there." I nod, accepting his assessment. Later, while we were inside getting drinks, he struck up a conversation with Paulo. He asked about where Paulo goes to school. The university? He suggested. No, Paulo said, the local community college. (At the time, Paulo was supporting himself, working and taking classes part time). "Oh. What are you taking?" Kim's husband asked. Paulo replied that he was taking office management. "Oh," Kim's husband offered again. At this point I was relieved that at least people were talking to Paulo at all, since I was worried he was feeling out of place. But then Kim's husband started talking about Brazil, and how Rio is the scariest place he's ever been. I said, "Scariest? Really?" trying to get him to qualify his statement and rethink what he was saying in front of Paulo, a native of Brazil. But, he said, "Yes," again, adamantly. "Great," I think. I was getting worried now, since Paulo already seemed a bit uncomfortable, and because I know from past experience that Kim's husband can sometimes say things that come off as pretty rude. "Well," he says to Paulo, "It's because I was wearing a suit. We were walking targets," he explains. "Did anything happen to you?" I asked. "No," he said, "but some of my friends were held up, and stuff like that," he replied, looking at Paulo. Then, suddenly sensing the awkwardness, he tried to repair the situation by saying that regardless of this, the mountains in Brazil were beautiful, and the ocean was also amazing. "Yeah," I thought, "let's just change the subject." A little later another guest asked Paulo why he was not eating. "Not hungry?" she says. Paolo smiled, and I explained that we just came from the fire hall potluck. "Oh, I was supposed to cook for that," the guest said, "but I didn't. Is that bad?" she asked, perhaps rhetorically. "What were you going to make?" I inquired, curious. "No," she says, implying that I misunderstood, "I never even called back," she clarified. Kim then also chimed in that she just gives money a couple times a year. She would rather just pay the money and not be bothered, she said. I nod to suggest that I heard her, but I am also thinking to myself, the purpose is not just the money but also the community and the socializing; the money is important to the firehouse, of course, but so is the sense of togetherness. As Kim and her guests continued to talk and enjoy each other's company, prompting another, different sense of togetherness, I began to understand that the community of Plainville is divided, and that I have experienced both sides of that divide in one evening, over the course of two dinners. Paulo talks to me about this later, and tells me that he noticed this too. "It was like Kim and her guests are too knowledgeable about food to go down to the fire hall and eat *that* food," Paulo said to me. "But," he protests, continuing, "the food wasn't really what everyone was there for, even though the people who attended were all enjoying it." At Kim's, food wasn't the only draw either, of course, but it was definitely given elevated status. We were drinking home-brewed beer and wine, eating the harvest of her garden, and even "talking food" (as foodies often do). It was clear to all present that Kim believes so much in how she produces her food, and that she considers it the "right thing to do." It is her moral duty, it seems, to feed her guests in the best possible way that she can. And her guests, through their praises of her labor, certainly appeared to appreciate this as well. Interestingly to me, however, none seemed to hold the feeling that showing up to the fire hall potluck, perhaps to share such a similarly home-made dish, might also be "the right thing to do." This was a sentiment that most of the potluck attendees would probably have agreed with, and yet it seemed to not occur to Kim or her guests. (To be fair, perhaps they had tried in past years, and their food had gone uneaten; I would not be surprised if this were the case.) I recalled that at several of our previous meetings, Kim had commented often about how hard it is to produce food in the way that she does, and how much work and personal dedication it takes. The implication, it seemed, was that this is what set her apart from the others in her community–those that don't take the time, or make the effort, to produce food in the same way. "Perhaps this was why she gave money, but did not think to attend the potluck," I thought to myself. "She disagrees with the manner in which the attendees produce their food." Towards the end of the evening, Kim talked more specifically about her labor and fore-thought in the meal preparation, delighting in the fact that the planning really begins with the seed catalogue in the spring. Another foodie in the group disagreed amicably, saying, "No, it actually begins with shoveling compost!" I smiled at the stereotypical foodie banter, but by this point Paulo was not talking much at all, and I was also feeling increasingly cognizant of that fact. As we lingered over desert, I tried to bring him in on a conversation about learning languages (something he is quite exceptional at), but I missed my opportunity when the conversation shifted. By the time we took off, Paulo left the evening in a noticeably lower mood than when we arrived at Kim's, being turned off by a general lack of connection to the partygoers, as well as the specifically disrespectful comments from Kim's husband. I left in a puzzled and uneasy mood as well, wondering how to make sense of the evening overall, and also how to convince Paolo to return with me again to future events in Plainville.

I begin with this lengthy excerpt from my field notes because it helps to depict the tensions that I experienced during my fieldwork in Nova Scotia, which I interpret to be largely class-based. This story is particularly interesting because it highlights how the "doing" of alternative food, in the form of gardening, cooking, and eating local produce, can and has become an elite practice, even (or especially) within a rural area like Nova Scotia. This phenomenon is one that I began to call the "gentrification" of simple eating, by which I mean that the practice of growing one's own food becomes taken up as a morally based leisure activity by the educated, upper (in relative terms) classes. This is not to say that those like Kim took the act of gardening lightly, as leisure, but rather that for those like Kim, gardening has become an individual choice that is driven by one's access to time, money, and most importantly a belief or faith that gardening is the "right" thing to do (for one's body, the community and the earth). In contrast to the "choice" of gardening stands, industrial food (the processed pies, cakes, and casseroles of the fire hall potluck) has taken the place of subsistence production in many poorer, rural areas like Plainville, Nova Scotia, because these foods are (now) cheap, easy to prepare, processed, and fast. As the two accounts above suggest, preference for local versus processed foods in Plainville has become a way to distinguish between the (relatively) elite and educated on the one hand (including myself as a researcher "from away" who also admittedly preferred Kim's food), and the less educated, rural folk of Plainville on the other.

Another important point to consider in this two-dinner story is that Kim is also a leader of the Plainville SGCP. Kim's sense of moral correctness in regard to her own gardening therefore directly translated into the way that she helped to define and coordinate the running of the school SGCP *as an enlightenment project*. Kim and other leaders of the Nova Scotia SGCP often discussed the program in terms of bringing knowledge and ideas to a community in which these qualities were perceived to be lacking or even entirely absent. Of course, as much scholarship in rural and urban identity has shown, this perception is consistent with how rural places are typically constructed as backwards and empty (see Ching and Creed 1997, Corbett 2007), in contrast to an educated and urban elite (including those who are "rural" only by choice). In the Plainville case study, a large part of this construction particularly revolved around how the SGCP leaders regarded the food traditions (or lack thereof) in the community:

> Traditionally people ate very basic, poor people food. Dried cod, green peas . . . there was no elaborate food tradition like in Italy or the southern US, just an inexpensive, rudimentary approach to cooking . . . [so here], it very easily transferred over to canned and industrial foods, and fast foods.

> (Leader, NS)

Because the tradition of food preparation in this rural area of Nova Scotia is a "basic" one, defined largely by what is available and cheap, some of the SGCP leaders interpreted this heritage as simply lacking or devoid of real value.

Elite cooking

In the Plainville SGCP, there was a particularly revealing story of the above dynamic that was repeated often in my interviews with leaders, teachers, parents, and students. Every year, the school holds a special event in which a locally famous male-identified chef is invited to the school to cook with the students. The chef arrives with his chef hat on and a car full of fancy equipment and spices.

The day proceeds in the chaotic fashion of an upscale restaurant: hectic, exciting, fast-paced and entertaining.

One year (prior to my arrival), parents were invited to this cooking event, in order to partake in the creations that the chef and their children had created. The scenario did not go as leaders and teachers had hoped. Some parents were noticeably turned off by the food, and refused to try it; it was too spicy, or unusual for their tastes. As one SGCP teacher told me, some parents even said "Ewww. What is in there? I am not going to eat that!" (Teacher, NS). The students were disappointed that their efforts were not recognized, and teachers were shocked and upset. The following year, the SGCP school committee decided that the parents would not be invited again; instead, the school decided to invite local politicians—people who they could count on to make the students feel proud: "The next year we invited people who we could be sure would appreciate the food, and validate the students' effort. Politicians. They were all very impressed, and that made the kids feel good" (Teacher, NS).

In an interview that I conducted with a SGCP leader regarding this event, we discussed the problem of the parents' lukewarm reception to the food. During the interview, I tried to get the leader's perspective on a variety of potential ways in which more members of the local community could be brought into the process of food production within SGCPs (perhaps through recipe sharing, for example), such that the negative (re)actions to SGCPs and feelings of judgment that were articulated by some parents I interviewed could be avoided. In response, the leader described the Plainville community as "unsophisticated" in terms of its food preference, and suggested that the chef's job during this event was to bring culinary refinement to the area, as a sort of labor of enlightenment. In his interpretation, there was little room for active participation in the SGCP on the part of local community members:

> If you were to have 20 people at [Plainville] send in recipes you would probably get 10 different recipes for mac and cheese . . . This is still a relatively white bread area, you know, with relatively unsophisticated palates . . . [your] idea of sending recipes in [as a way to include the community] is interesting, but I don't think you would come up with a wide range of things, nobody is going to send in a bok choy recipe . . . it's problematic; [the chef] was brought in to show them that there is diversity, rather than the same old same old. [They] did a soup and put everything in the blender, and, my God, it shocked everyone . . . parsnips and apples [in one soup] . . . So [the chef's] role is to widen their variety . . . so the parents themselves, who really don't have any idea how to cook beyond the box . . . [he can help them] to create healthy food and also develop tastes for their kids.
>
> (Leader, NS)

Here, the leader suggests that recipe sharing is not really a possibility because those in the local community lack the correct food attachments (preferring mac and cheese to bok choy), and thus the correct food knowledge (not knowing how to make anything beyond the box) to be able to share appropriate recipes for the event. He further clarified his position by explaining that those in Plainville who eat processed foods and shop at big supermarkets rather than farmers' markets are "lemmings" and that "[they] are being led by [their] noses," adding "I don't want to sound elitist, [there are] rich lemmings [too]" (Leader, NS). In such comments, there is an implication that the local community members of Plainville— those who are not highly educated, or who have not lived elsewhere and have not experienced other culinary traditions—are not smart enough to make the "right" food decisions, and therefore cannot effectively participate in a culinary event.

Changing (some of) the parents

This way of thinking, in turn, has an impact on what such SGCP leaders imagine the overall goal of the program to be. In short, the goal becomes one of getting to the kids so as to change the supposedly inadequate food behaviors of the parents. This goal, however, seems under-realized at best:

> *Leader:* I guess the one negative that I hear is that it hasn't really [changed parents] to the extent that we would have liked. It's hard to change parents' habits. And it's hard to change how they've been brought up and what they are used to and even though we try, and students do go home and say you know let's try this, I'm sure that sometimes the parents will cook it for their children but then still not . . . refuse to eat it themselves. [Or] so I think . . . because we've had comments made about different functions we've had and the parents and refusing to eat.
>
> *JHC:* Well, I heard about that at the event that you were talking about with the chef.
>
> *Leader:* Yeah and so in terms of getting that message to parents that perhaps they might want to try some of it . . . I would say that if the parents aren't going to eat it they are not going to cook it, most of the time. So the carry over may not be as strong at home yet.
>
> (Leader, NS)

In this explanation, it is clear that the role of the school is imagined to be one of enlightenment and that the SGCP can be considered effective if they succeed in the mission to "change parents' habits." Yet, it is not *all* parents in this imagining that need changing; it is just the ones whose eating preferences do not fall in line with the school's notions of healthy eating. Such preferences, however, are not randomly distributed across the population of Plainville. In fact, according to some of my interviewees, two fairly distinct groups of people have emerged in the community, with two distinct food identities: the locals (who now typically eat processed, industrial foods, as I experienced at the potluck) and those "from away" (who prefer alternative food and shopping at farmers markets, as I experienced at Kim's house). As a mother "from away" described to me this dynamic:

> There is a large community feel [here], but there is an alternative community consisting of people from away, and then a local community. They don't mix much. I think the alternative community has to put more effort into integrating and bringing along the local people, to a festival or a market or whatever.
>
> (Mother, NS)

A SGCP teacher also articulated a similar split, offering: "There are people in this community that are always willing to try new things. They are educated, and well traveled, and open to new ideas. And then there are those that aren't" (Teacher, NS).

This social grouping of Plainville families into two distinct categories also translated into how some of the SGCP students experienced food at school. For example, during a peer interviewing exercise in which students interviewed each other about food habits, I witnessed this dynamic between two students, one from each "side" of the divide. The one student was obviously well traveled, and knew a lot about other food cultures beyond rural Nova Scotia. He asserted that *none* of the school food was actually good enough for him, even the healthier food, because it was *all* too simple and plain. The other student was from the local community and had a family who had been there for generations. He was indifferent about the school lunches and the SGCP, but said little because the other boy dominated the conversation. I wrote about this incident in my field notes.

Field notes, September 2007, NS

This one kid was interviewing another and acting very superior, saying he didn't like the food here because it wasn't from other countries. He had obviously had a lot of food experiences, and his superiority came in the form of him wanting to differentiate himself from his classmates and even Canada. He said that he had been to 10 different countries, and said that everyone here [in Plainville] drinks at least one pop a day, besides him.

This student "from away" represents somewhat of an anomaly in the school, in terms of how adamantly he refused school food. Still, for the most part, what was cooked and served for school lunches did seem to fall somewhere in between the food habits/preferences of the two "distinct" camps. That is, while those from away (mostly parents) had suggestions for how to make SGCP food more healthy, or more alternative, those from the local community tended to want some flexibility in what the school served, vacillating between "healthy" food and "junk" food. For example, one "local" mother asked me, "What did you have growing up? White bread, right? So why can't our kids just have white bread some of the time?" (Mother, NS). In contrast, when I mentioned to a mother "from away" that the school sometimes uses yogurt with Splenda in it, she gasped: "No! That's sucralose! They feed that to the kids?" (Mother, NS). Thus, when I asked "local" students how the food they learned to make in the SGCP compared to what they made at home, the typical response was simply, "It's different." Often these differences revolved around the use of processed, easy-to-prepare foods:

> *Student Z*: Different.
>
> (Student, NS)

> *Student Y*: Different because they cook stuff like Side-Kick Noodles [processed].
> (Student, NS)

> *Student X*: Different.
>
> (Student, NS)

> *Student W*: I have like fish sticks at home. And we don't have to do work at home.
> (Student, NS)

> *Student V*: I eat, ummmm, different foods. Like I eat Kraft Dinner.
> (Student, NS)

Given the conspicuousness of these differences, the lack of attention that the SGCP leaders in Nova Scotia gave to issues of class-based food identity was surprising. This is especially true considering the recent academic and practitioner work that has been done in the region to expose issues of poverty and food security (e.g. Williams 2002). Despite this, however, when I brought up issues of purchasing power or geographic availability during my interviews, the interviewees who fell into a more "elite" class status sometimes refused to deal with such issues, insisting that ultimately it came down to setting individual priorities and making appropriate choices. "Healthier food is cheaper anyway;" one leader in Nova Scotia told me, "you just have to take the time to prepare it, and you have to know how" (Leader, NS). As another leader in Nova Scotia insisted to me repeatedly, "Processed foods actually cost more; it is cheaper to eat healthily" (Leader, NS), adding that, "education could cure most of the health problems in the area" (Leader, NS).

These sentiments were also particularly surprising considering that many other students and parents from lower social class backgrounds expressed openly to me that access to fresh, local ingredients outside of school was a financial issue for their families. While I never heard the SGCP leaders repeat their beliefs directly to any SGCP student or parent, undoubtedly their persistence in blaming individual behaviors fed into the structuring of the SGCP as an enlightenment project, and thus also to the feelings of judgment that parents often articulated or alluded to during interviews. Because the leaders interpreted the local community as "lacking" any significant food tradition, their own familial attachments to food (be they processed or local) were undermined, rendering parents passive in the SGCP's process of teaching their children how to eat "better."

Rurality, resourcefulness, and scrapper identities

> So, you have [this dilemma] where a kid feels bad because he goes home
> and is like, well we don't set the table, or have a table cloth, or flowers. But
> I think Alice [Waters'] dream is that they go pick some daisies or do the scrapper
> thing, which is what I did. Eventually you are like, ok, how can I get this
> without paying for it? Like real peasant food . . . and you have the skills [from
> the SGCP] to turn it into something delicious. People who are motivated will
> do that, and others will just be bitter.
>
> (Activist, CA)

Although some of the leaders of the Nova Scotia SGCP were inattentive to issues
of class-based food identity, and were thus implicated in the reproduction of class-
based hierarchy within the local community, I also came to recognize that the
daily operation of the SGCP itself also allows for moments of hierarchical disruption.
Once again, this disruptive role is related to the opportunities that the SGCPs
provide to students and parents to relate to alternative food/practices that they
otherwise might not encounter. As I discuss in Chapter 3, by expanding people's
relational networks to food and food practice, SGCPs can also help bring about
new visceral horizons that interrupt and unsettle the status quo of food-based
relating.

Cooking simple

As the extract above from an activist in Berkeley points out, one way that SGCPs
do this is simply through encouraging the development of new skills. Indeed, by
providing students with a connection to the daily, lived space of the SGCP kitchen
and garden, the Plainville SGCP teaches students a great deal about how to grow
and prepare home-cooked meals. Therefore, although parents did express feelings
of judgment or critique in regard to SGCPs, many parents, and particularly some
mothers, also acknowledged the importance of cooking and gardening as skills that
were rapidly being lost:

> I think it's great, it's good for the kids to see food grown, planting seeds and
> actually seeing the lettuce and the carrots. Because I find now, not so much
> everyone has gardens anymore, we don't have one, and it's wonderful for
> them to make the food with it.
>
> (Mother, NS)

> There are so many kids whose parents [both] work now, so supper is usually
> a rushed fare and it is so good to see things being prepared, not coming out
> of a can or a package. And we don't eat a lot of that food in this house, but
> for those who do, it opens their whole horizons to what can be done.
>
> (Mother, NS)

It is also important to note that in Nova Scotia, few if any of the meals that SGCP students cook on a regular basis are "fancy" relative to the meal that they cook once a year with the professional chef. Indeed, the "simple, no–nonsense" meals that they normally make in the kitchen with Ms. Dora include "salads, veggie wraps, pizza, potato wedges, and vegetable stews like hodge-podge, which is [actually] a local culinary tradition of rural Nova Scotia" (Field notes, September 2007, NS). Although some parents were skeptical to the overall changes in the lunch menu (from hot dogs and chicken nuggets to such vegetarian fare), the students themselves had mixed reactions about the changes in their school menu—viewing them as "important" and "good" overall, while at the same time offering (me) suggestions for how the daily selection could be more "balanced between healthy and junk [or processed] foods" (Student, NS). Further, in the daily, lived space of the SGCP itself, I sensed little of the tension that was expressed by some "local" SGCP parents, and to the contrary, I witnessed much interest and excitement on the part of the students towards growing and cooking if not actually eating the food. As a leader of the school's PTA explained to me, many of the same parents who were initially put off eventually came to admit that their children did often enjoy the (vegetarian) food, and also the opportunity to be in the garden.

The oft-repeated story of the "problematic" chef event is also more complicated and contradictory than a simple class conflict analysis would admit. Many students enjoyed meeting and working with a trained chef, and were excited to be working alongside him. Although not all students and parents like what they cooked, the experience itself was a positive one for many of them, and many articulated as much to me during interviews. While some parents clearly interpreted the chef event as elitist, like I initially did, the children themselves seemed to have mixed experiences of the event, at least according to those with whom I spoke. They enjoyed the chance to do something different during school hours, and they were able to engage physically in a hands-on event with a chef that they otherwise would not have encountered. Most students considered themselves lucky to have this opportunity.

Thus, the chef event also needs to be considered in terms of its potential for expanding students' horizons. For example, it is certainly possible that by meeting this chef, and spending an afternoon cooking with him, some students were then able to recognize that cooking might be something to which they could aspire as a profession. Of course, one might ask whether this also reinforces the importance of cooking as a (male-dominated) money-making occupation over other forms of cooking (as in women-centered care work, domestic labor, etc.). However, an alternative interpretation could also be used to re-examine the event—one that recognizes it as simultaneously an extension of the students' food networks and possibilities. While initially I read the event as more of a class struggle of sorts, the chef himself also gave me this explanation (as one of several) of what he was doing there: "The majority of kids in this area do not have access to a diversity of experiences, so [meeting a real chef] is a treat for them. It is one more opportunity to know what you might do in your life" (Leader, NS).

Moreover, beyond enjoyment, awe, and other positive emotions, as some interviewees suggested, additional benefits may even arise from conflict itself. The negative reactions of some parents to the food did not necessarily change all students' experiences of the day, but it surely made at least some of them feel confused or invalidated. Nevertheless, some of the teachers of Plainville School suggested to me that they were not worried about such conflicts, because it is "only through conflict that you can make change" (Teacher, NS). It is clear that some teachers and leaders were upset enough at the feelings of negativity that this conflict produced to take steps to avoid the conflict in the future (by inviting politicians the following year). However, as Elspeth Probyn points out in her own work with food (2000), in avoiding such negative visceral responses, we lose the opportunity to understand each other through our differences, and to begin to recognize and dismantle the hierarchies that such food conflict reveals. By allowing only positive (re)actions to food, we could say that avoidance of conflict may actually conceal the diversity that is present within all food relationships, and relinquish the chance to engage students and parents in meaningful dialogue about food. Thus, in regard to the chef event, negative reactions are not necessarily "bad."

Healthy networks and possibilities

Other parents suggested to me that they were grateful that they could be sure that their kids were getting at least one healthy meal a day, even if they could not provide that meal at home. Many of the teachers and administrators of Plainville School were also aware of this function, and thus they saw the SGCP as providing students with a daily material connection to healthy food that they did not already have, which undoubtedly encourages new visceral attachments. This understanding mirrors the discussion of nonwhite food activists that I highlight in Chapter 5 in regard to race and food in certain areas of Berkeley, where economic and geographic connections to fresh fruits and vegetables were minimal in poor communities of color. In the physical act of relating, new possibilities can emerge:

> Again it goes back to, I think, that time and money play a role in you know, let's get something frozen we can put it in the microwave that kind of thing. And I understand that and I think those kids who come from situations where the parents work late. Very often they might be getting supper on their own and it is easier to open up a can of something; you know those kids are really benefiting because they might not have ever eaten turnips. I think that there are a large percentage of students who are eating the turnip who never would have—either because they never had it or they didn't think they liked it or whatever. But the whole experience . . . then they realize that it's not so bad. So that whole door to all kinds of different foods has been opened that maybe wouldn't have been before.
>
> (Leader, NS)

Beyond providing physical access to healthy foods, another way that the Plainville SGCP promotes resistance to class-based hierarchy is by inspiring members of the rural community to identify with healthy, alternative food through re-establishing cultural roots with the area's subsistence past. As several "local" parents pointed out during interviews, a rural area like theirs is the perfect place to establish such a school garden because of the long tradition of family-based, small-scale agriculture in the region. Some parents implied that teaching students how to garden was a way to honor this tradition and to reclaim the importance of rural community, rather than educating students to want to leave rural areas and livelihoods (Corbett 2007). Certainly the concept of local eating—even in Alice Waters' ideology—rests in a kind of nostalgia for the rural ideal, in which life becomes based in simple, modest practices that do not require much elaboration or fanfare (see Waters' book *The Art of Simple Food*, 2007a). In this sense, the philosophy of SGCPs and the cultural traditions of the "locals" in the community are not necessarily that far apart. As two mothers offered:

> Being in a rural area, we have the space for [such] gardens. And [in this type of community] you know everyone, so if you don't know how to farm, you know someone who does.
>
> (Mother, NS)

> [It's relevant to us because] healthy eating is really about getting back to the basics, getting away from preprocessed foods and fast foods, and living off the land again.
>
> (Mother, NS)

In these mothers' understandings, the SGCP can help to legitimize the importance of rural livelihoods, and can encourage students and parents to be proud of their rural identities. The SGCP can recognize rural communities as places out of which knowledge emerges, rather than as places where knowledge is lacking. Through my discussions with these and other mothers, I came to recognize that parents play a crucial role in the functioning and relative "success" of SGCPs, although their role is also often under-regarded. Parents' (re)actions to SGCPs matter in significant ways, and their involvement with SGCPs impacts how effective the programs are at bringing change (of any kind) to the school and community. As one father in Nova Scotia commented, "The [SGCP] just has to realize the real change is gonna come *with the parents*; how you get the parents involved is really important" (Father, NS).

Following the cue of this parent, as well as my own growing discomfort with the class-based tensions in the Nova Scotia case study site, I therefore decided to use my remaining interviews with parents and leaders as opportunities to also explore how parents *could be* involved more effectively. In this sense, my interviews themselves also became (partial) spaces of hierarchical disruption or resistance to bounded class-based identities, although my interviews with parents were generally more successful in this regard than were my interviews with leaders. Several parents, both local and "from away," had important and interesting ideas about how to change the contentious dynamic:

They should make good hearty food, not fancy and gourmet [with the chef] . . . I think in the city where people are used to that, they would like it, [but] people around here, we have a very particular way of preparing food.

(Mother, NS)

What if the grandmothers of the community could come together with [the chef] and prepare a menu that would appeal to the community? I am just thinking of these suppers that the local churches and firehouses do, with baked beans and brown bread and hodge-podge. I mean, as simple as it sounds, squash soup is not something that these people would eat . . . the idea of [the chef] coming out here to [the school], sounds immediately wonderful to me, and to people who know about him, but as I think about it more, I realize it may not be the best approach. I am sure there are a variety of other ways to make it more comfy . . . but one way I think is to actually have the grandmas come in and offer what they have, recipes, cooking experience, or maybe both, start with grandmas and then have someone like [the chef].

(Mother, NS)

The fact is that women do more of the cooking if not all of the cooking in this area culturally. [Women, and especially grandmothers, hold the] cooking experience and stories that go with it . . . so, they could bring that. I think the grandmother thing would be good . . . [the chef] is a very down to earth kind of a guy, very sensible to and sensitive to these ideas, and [when] he makes a potato leek soup it is something . . . with some coriander . . . it's something that some people have not had in their lives.

(Father, NS)

There was a sense among some of the parents that I interviewed that it was very important that the SGCP meet the parents "where they are coming from" in regard to their own food traditions, if they want to bring parents "on board" (Mother, NS). In this sense, the question to be asked by the Plainville SGCP is not how can we convince the (local, less "educated") parents to think/act differently (or like us), but rather how can we include all parents *as already* thoughtful and active participants? In other words, how can we legitimize all parents' food identities in a way that allows for and even encourages a diversity of opinions and practices within SGCPs? Moreover, as the last two comments regarding the involvement of grandmothers suggest, this re-framing of the parent question is important not only for disrupting class-based social hierarchy, but also for disrupting a gender-based hierarchy as well (the subject of Chapter 6). Indeed, it is problematic and counterproductive that, in the Nova Scotia SGCP, the sole example offered to the students of a "successful," professional chef is also a person who is male-identified. This representation of what it means to be "successful" (economically and socially) stands in stark hierarchical contrast to both Ms. Dora as the lower-class "lunch lady" figure-type, and also to the myriad mothers and grandmothers who (according to my interviewees) constitute most of the nonpaid cooking population of the

Plainville community. To deny that these women hold important knowledge about how and what to cook is to undervalue the work of these women as mothers, wives and members of the local community.

Concluding thoughts

One of the frequently articulated purposes of SGCPs is that they are meant especially to help kids that are particularly "at risk." This is also often confirmed by the funding sources of SGCPs, which are often intended to support school programs that nutritionally target disenfranchised youth—that is, students from poor and/or minority communities who have higher rates of dietary disease (USDA 2014). But, as this chapter's examples illustrate, we also need to ask the question, in what ways do these children actually need "help"? For example, can we encourage programs that help children to negotiate tricky food dilemmas? Can we encourage them to find their agency within the broader, inequitable food system? Certainly it does not help lower-class children when they or their parents feel judged about the way that they eat at home, while better off families simultaneously enjoy having their eating habits legitimized. But, as we have seen, SGCPs are involved in not just reproducing but also resisting class-based hierarchies. Therefore, as class clashes and inequities become a more explicit concern of SGCPs, the programs can begin to "target" children's actual needs, and encourage more resistances to class-based hierarchies.

5 It's not just about the collard greens

Central School's garden in Berkeley grows both kale and collard greens; the kitchen serves both tofu and tamales. Nevertheless, the racial codes and categories that shape different students' experiences of their SGCP go far beyond the actual foods themselves. Despite the diversity of choices in both garden and kitchen classrooms, students of color in the Berkeley middle school continually insisted to me that the food they were learning how to eat was *white* food. "But it's not just the food," explained one student, "it's the whole experience . . . the way we do things, and the [reasons we do them]." This chapter explores how racial identity and food practice intersect in complex ways, and how what is compelling to some can feel irrelevant or even disrespectful to others. Importantly, however, it also explores how the physical body intersects with food identities in (sometimes) unpredictable ways, creating outcomes that can be both inconsistent and transgressive. In the sections that follow, I discuss both the ways that racial hierarchy is reaffirmed and resisted within SGCPs. First, however, I want to examine the meaning and matter of racial identity in a bit more depth.

Pockets of subjectivity in race and food identity

> It is hard to overcome students' pre-set feelings to certain foods. It is hard, you can only do so much.
>
> (Leader, NS)

> I think children are clean slates when they are born. As adults, we impress upon them so much, so [certain tastes] are set in them. But, it doesn't matter what age you are, 10 or 50, the brain is still pliable.
>
> (Mother, CA)

The research experiences that this chapter explores highlight the intersection between social identity and food, examining race in particular. I begin with the above quotations because I want to discuss food identities in terms of their viscerality—that is, the ways in which food practices and tastes become bodily habits of being and feeling, from which we practice and perpetuate our social identities. Students come into the SGCP classrooms with a variety of different food identities, which are not necessarily "set" or "fixed" but certainly habituated to some extent into their daily life practices. Building on previous chapters, we can think of food identities as the visceral intersection of social subjectivities and food practices—that which propels judgments like "This is my kind of food," or "I don't eat this way," (although such judgments are not always consciously articulated as such). These food identities are at once discursive and material. They are defined based on already established, and always intersecting, social categories: black, white, middle class, woman, youth, etc., but they are also always *more than* these categories in that they are taken up and reproduced through bodily practices that do not always remain within the boundaries of one discursive grouping.

SGCPs provide a space for students to both reproduce and resist the existing boundaries of food-based identities, particularly in regard to race, class, gender, and age. My contention in this chapter is that SGCPs do their racial work within and through such food identities, and that these food identities can both be limiting and enabling to progressive social change on issues of race and racism within SGCPs. Rather than promoting an understanding that eating good food together will *overcome* our differences, as SGCP leaders often suggested, I therefore contend that we need to begin to attend to how racial difference itself is a crucial (and visceral) proponent in even bringing us to the table. Talking about racial identities in visceral terms, though, is a tricky task, particularly because it is necessary to make use of discursive categories at the same time as we allow social identities to spill beyond these boundaries and be(come) more than these categories. In regard to issues of race and geographic identity, for example, sociologist France Winddance Twine (1996) highlights black women's concurrent need to recognize and celebrate the black community as a distinct(ly oppressed) group, while also working to break down fixed assumptions about what constitutes blackness to begin with. What then, if we wanted to call for better access to healthy foods that are culturally relevant for a black community? How would we define what healthy eating *is* for a black community, while holding the meaning of both "healthy" and "black" in question?

While these conflicts may sound unanswerable and perhaps overly philosophical, perhaps one strategy is to try to practice, rather than solve, the problem. That is, perhaps we can begin by consciously attempting to live our lives in the midst of such tension—as though these modes of identification are significant but not enduring parts of our daily, bodily existence. In this light, feminist geographer Audrey Kobayashi advocates recognizing "all forms of social identification (especially race and gender) as temporary and dissolvable political coalitions rather than fixed groups" (1994, 78). Instead of attempting to overcome or ignore social labels, she insists, quoting Seamus Deane who says that: "The divisions of [the past] must be lived through in the present. It is therefore necessary to sustain commitment to them under the aegis of irony. Otherwise the oppressive conditions they bespeak will merely be reproduced" (Kobayashi 1994, 78).

This strategy is notably similar to Gilles Deleuze's and Felix Guattari's notion of "becoming indeterminate" (1987), though the phrase might not initially sound parallel. In Deleuze's and Guattari's discussion of such dilemmas of language, they explain that they want to maintain a radical openness to the possibilities of (material) human becoming—to what we are to be in the future. In this sense, Deleuze and Guattari seem to eschew labels altogether by advocating an ideal "goal" of total indeterminacy, where terms like "black" or "girl" or "healthy" cease to define or distinguish us from each other. Like Kobayashi, however, Deleuze and Guattari do not outright dismiss these labels. They also recognize that because we exist in the midst of already established social hierarchies, in this "open" process of material becoming it is necessary to retain "pockets of subjectivity"—little bits of (bodily) faith in the labels themselves—in order to be able to (quite literally) feel our way forward (Grosz 1994, 171). In other words, Deleuze and Guattari realize that without any sort of temporary sense of "blackness" or "healthiness," for example, we would be left unable to act at all—we would be rendered socially ineffective. In a way, then, our political agency comes in and through the (temporary) fixity of these labels—and simultaneously through our (bodily) ability to throw them into question. In this sense, the *becoming* of "becoming indeterminate" is perhaps the more important word of the phrase, for it is the active, visceral practice of "living through" the impermanence that matters more than the imagined (and unattainable) end solution.

Another important point to consider here is that the very (material) practice of indeterminacy or impermanence is itself always already influenced by past hierarchies and social orderings. Thus the ability of a person or group to practice these "pockets of subjectivity" *as temporary* is itself determined by their already (somewhat) structured place in society. For example, as feminist geographer Mona Domosh (1997) reminds us, identities are "fluid and never fixed, although dominant powers *structure the terms and extent of that fluidity*" (85, emphasis added). Therefore, it is important to recognize that those in a place of racial privilege in our social world are often much more likely to be able to engage in the fluid practice of indeterminacy or impermanence than those in positions of relative disenfranchisement. While I can quite easily practice impermanence in my academic writing and my interpretation of text, it is not my access to such practices that ultimately matter for SGCPs. In this light, it is not enough to simply advocate the practice itself, we must also work to make this practice accessible to disenfranchised "others." One of the questions propelling this chapter is, therefore, do the SGCPs make such a practice accessible to all students?

As I discuss racial identity and food in this chapter, I try to take seriously these responsibilities both in my own writing, and in terms of how I analyze the work of the SGCPs in question. I use terms like "black," "middle class," "healthy," and "woman," frequently in my writing, reflecting the language that my research participants used during interviews and classroom activities. I use these terms with the recognition and insistence that they are significant ways to distinguish between social groups that have distinct histories and knowledges, and also that these labels are constantly and actively being shifted, altered, and resisted from both within and beyond the bounds of their current meaning. It is important to also mention once again that feminist scholars have traditionally been interested in the *intersectionality* of race, class, and gender—what has become considered the feminist triad—as it is frequently used as a way to analyze and assess across lines of social difference (Valentine 2007). (I also discuss age because working with youth made me recognize that age is another important food-based identity marker, one that many of my student research participants discussed and practiced frequently.) As I discussed in previous chapters, while I separate these four in this book for the purposes of illustration and clarity, I also recognize that the four identities are not mutually exclusive but rather co-constituting. Racially based food identities, for example, are also often formed within the context of class differences, and gender-based food identities are further distinguished by race, class, or age differences. Below I discuss these intersections as much as possible, while also maintaining the separate categories in order to be able to demonstrate how the work of SGCPs both reinforces and resists the boundedness of each category.

Race and food in Central School

The intersection of racial/ethnic identity and food is a topic that became apparent during my research in Berkeley, CA. Because the population of the area surrounding Plainville School was predominantly white (98 percent), racial/ethnic identity as a form of social difference within SGCP spaces was not apparent in my Nova Scotia case study. Central School in California, however, was notably more diverse. According to school estimates, approximately 50 percent of the schools' students are African American, 35 percent are white, and 15 percent are Asian. Also, 15 percent of students are Latino (of any race). Though not calculated by the school, these categories also include many students with a multi-racial or ethnic identity.

Though Central School was racially and ethnically diverse, much of the student-led discussion around racial identity and food in my research was in reference to two racial identities: black and white. To a lesser extent, some students also referred to Latinos as a distinct ethnic group, while there was little to no mention of an Asian identity. This is perhaps partly due to the fact that most students within the SGCP classrooms identified as either white or black, and thus the majority of students with whom I spoke discussed racial identity along such lines. At the same time, however, students also seemed to use the term black to refer to a general "other" to white—the dominant category that was perceived by many nonwhite students to color the space of the SGCP classrooms and the school as a whole. In this sense, the label black also seemed to encompass a variety of (mixed) racial identities, and co-existed along with Latino as an "other" to white.

In addition, it is important to note that "whiteness" as a label or identity marker was not discussed or used as such by most of the white students or teachers who I interviewed or came to know. For this reason, while I present a great deal of data (below) on how "black" students experience and construct their identity vis-à-vis SGCP food, I have little to no direct discussion on how white students (or teachers) negotiate their own racialized food identities *as white*. The reasons for this are perhaps various, but one particular explanation is that white students and teachers do not recognize their food identities in directly racial terms because white identity is often experienced as racially neutral (McIntosh 1989, Twine 1996). As Twine illustrates in her own work on whiteness and black identity, in many places within our contemporary North American society, whiteness is invisible insofar as those associated with a white identity view their own racial categorization as culturally and politically neutral (even given, and perhaps because of, the hegemonic character of whiteness). In this sense, white students and teachers would be less inclined to experience their attachment to food in terms of race, whereas for "black" students, and indeed all students of color, race would be a more salient marker of food-based identity and difference.

The two sections below discuss the intersection of racial identity and food practice, drawing from interviews mostly with students, and also from my participant observation experiences within the SGCP classrooms. I first discuss ways in which the fixed racial categories of "black" and "white" seem to be confirmed and reproduced through students' experiences of SGCPs, such that two categories of food emerge: white food and black food. In the section that follows, I then move on to discuss how the bringing together of a racially diverse group of students within SGCP spaces encourages what I call "boundary events" (after Twine 1996)— moments in which the boundedness of such racial categories as black and white is called into question.

White food and black food

> If we left it up to self-selection, it would be a group of 20 white girls. So we really attempt to recruit kids to get more diversity.
>
> (Leader, CA)

The above quotation is in reference to a summer program that is run every year at another SGCP within the Berkeley school system. While this comment does not refer to Central School particularly, it is telling in terms of who such programs in general tend to attract within the Berkeley community. If SGCPs are meant to be relevant beyond racial differences, then we need to ask, why is it that most of the children who would attend such a summer program are white (and female-identified)? Given that the alternative food movement at large has been noted and critiqued for its whiteness, it is important that we examine how SGCPs are also constructed around such racial lines. Do nonwhite children see SGCPs as relevant to their own lives? Is the motivation to eat healthy alternatives something that is coming from *them*—from their experiences and attachments? And what would such self-determined motivation actually mean or entail?

Unlike the summer program discussed above, during the school year at Central School the SGCP classrooms are racially diverse. The racial breakdown of the classrooms is roughly representative of the overall percentages of the school at large. For the 6th graders who take the class as a requirement, this needs no explanation. However, for the 7th and 8th grade elective class, it is important to understand why this is the case. The reasons for this diversity are somewhat contradictory. To begin, teachers and administrators recruit students from a variety of backgrounds to participate in SGCP classrooms. There is a concerted attempt to bring a diverse group of students together within the cooking and gardening classrooms. This attempt is meant to encourage students to share ideas and information with each other, and to engage in a process of peer knowledge production. According to student interviewees, however, recruitment takes place in the reverse direction from the summer school situation—that is, teachers often recruit white students to partake in the SGCP classes. In a group interview with three white girls, we discussed this issue:

> *Student A*: Our English teacher recommended us to publishing but some people didn't get into it because they wanted to make [the classrooms] more diverse; so now there are people in [publishing] who don't want to write, and people in the cooking room that don't want to [be there].
>
> (Student, CA)

> *JHC*: So did a lot of the African American kids want to get into cooking, and not publishing? . . . Is publishing considered more educational or academic?
> *Student A*: Yeah, well a lot of kids enjoy eating and cooking, if you want a more hands-on type of class. But it is weird because it is true that more of the kids getting 4.0s are white, it is a fact; but then [teachers] recommend people [for publishing] and then [the administration says] no you can't go, sorry, because we need more balance . . . It's [related to how] the superintendent talked about the gap in achievement [between white and black kids] and how that [needs to be dealt with].
>
> (Student, CA)

This discussion illustrates that at the 7th and 8th grade level, white students were the ones actually recruited into the SGCP classroom in order to make it more "balanced," something that my discussions with other SGCP students and teachers also confirmed. Using this information, we might conclude that the SGCP is more attractive to black students than white—the reverse of the summer school situation discussed above. This conclusion, however, leaves out the question of publishing; why are more white students attracted to publishing (and why, more broadly, are more white students getting 4.0s)? Further, how has the choice between publishing and cooking come to be racialized, in ways that are clearly linked (in the student's description above) to the broader systems through which academic achievement is defined and conferred? These conditions would suggest that it is not so much that students of color are more comfortable in SGCP classrooms than white students, but rather that the students' levels of comfort in either class are strongly tied to systems of academic achievement that are deeply racialized.

Under these circumstances, then, while black students seem more inclined to sign up for cooking and gardening, this does not necessarily indicate that they perceive their SGCP classes as relevant to their lives or as consistent with their food-based identities. In fact, later on in my interview with these same three students, I asked the girls about a term that I had heard students of color use to describe SGCP food—the term "white food." They responded:

> *Student A*: Yeah, we heard that . . . if you look around Berkeley you will see that the people who are all about organic happen to be white hippies, not like African Americans or Latinos.
>
> (Student, CA)

> *JHC*: Why do you think this is?
> *Student A*: I don't know, it is just what they are used to, what they grew up with.
>
> (Student, CA)

> *Student B*: Yeah, I think if you were walking in a farmers market, I mean no one thinks it is a bad thing, but you would notice if you saw a bunch of African American kids. It is not the norm.
>
> (Student, CA)

> *JHC*: Do you think this also applies to the [SGCP]?
> *Student B*: I think so. If you go around and look at people who buy nachos at 7–11 it is the black kids, so sometimes they are less affected by the cooking, because they are like, whatever, we will just eat chicken. It is one of those things, like how you notice a lot of the super achieving kids tend to be white.
>
> (Student, CA)

In addition to these students' comments, white parents often articulated similar assessments of "normalcy" in terms of food tastes and racial/ethnic identity, some-times with a tinge of blame:

> A perfect example is Picante, that Mexican Restaurant with Alice Waters' style influence. I love that place, and the food is good, but you know, it is not authentic [because the cooks have] ditched lard and heavy, bad ingredients. So, Mexicans won't eat there—it is not greasy enough. Their taste buds are locked into this instant gratification thing, not a nutrition thing at all.
>
> (Father, CA)

During my discussions with students, I heard the terms "black food" and "white food" used often. Generally "white food" referred to the types of foods that students were learning how to cook in SGCPs—vegetarian, low in fat, lacking in animal products, organic, unrefined, and whole foods. This is notable especially because in terms of culinary diversity, the recipes themselves reflected a wide range of cultural and ethnic traditions: Indian, Chinese, Mexican, European, American, Southern/Soul, etc. However, particularly because this food was associated with the labels "healthy" and "organic," what was cooked in the SGCP was perceived and discussed by many students of color as "white food." When I asked some of the students of color that I was interviewing whether the food they cooked in the SGCP class was like the food they ate at home, their responses indicated that there were significant differences:

JHC: Is the food we cook in here similar to what you eat at home?
Student C: Pssssh, no.

(Student, CA)

Student D: No.

(Student, CA)

Student E: No.

(Student, CA)

Student D: Some stuff I eat at home.

(Student, CA)

Student C: Yeah, some of it. Like the green beans.

(Student, CA)

JHC: So is the food here similar to what you eat at home?
Student F: Nope.

(Student, CA)

Student G: No.

(Student, CA)

JHC: What is different?
Student G: The food over here is nasty, we be eating like, I be eating tacos, hamburgers.

(Student, CA)

Student F: Jack in the Box.

(Student, CA)

Student G: Jack in the Box, I be eating spaghetti and meatballs. Yup.

(Student, CA)

JHC: So what would make this program more important to you?
Student F: Make better food. Cause like if the food wasn't . . . I mean, they make smoothies, and they don't got any of the right stuff. They gotta have soy milk, they can't have regular milk. When we want to use

sugar, they gotta have fake sugar. Like we want the white sugar, like just plain.

(Student, CA)

Student G: We can make a deal with them, like they have some nutritious food, and then some good food, and then some nutritious food.

(Student, CA)

Differences in food at school and home in my interviews often seemed to revolve around the lack of meat in the SGCP classroom, as I have noted, but the comparison also took other dichotomous forms: Slow Food versus fast food, "fake" sugar (rapadura or turbinado) versus "plain" white sugar, soy milk versus cow milk. However, these differences are more than just a matter of biophysical taste, and more than simply what these students are used to. They are linked in important ways to these students' racial identities, and are indeed experienced as bio-social. As one of the above students went on to tell me during the interview: "I feel like the teachers are trying to make us be someone we don't want to be" (Student, CA). In the latter of the above interviews, the two male-identified black students with whom I was talking went on to particularly discuss how race is associated with this dynamic:

Student F: Like black people is different from white people. You guys eat different food. Not saying you, but white people. They eat all that healthy food and stuff. We eat a lot of junk food. And they are always trying to control us.

(Student, CA)

JHC: White people try to control you?
Student F: Yeah you think we too hyper.

(Student, CA)

Student G: And black people eat different food from white people, like we even eat different food from Mexicans.

(Student, CA)

Student F: You guys make a lot of healthy food that is not good.

(Student, CA)

JHC: So is the food you eat here white people food?
Student G: Yeah.

(Student, CA)

Student F: We never have black people food . . . like ribs and burgers and fried chicken, Chinese food.

(Student, CA)

Student G: When black people try to get into the school district white people don't let us because I think you guys don't want to let us in because the kind of food we eat, we might try to change the school board and get better food.

(Student, CA)

Student F: Like we could go to Jack in the Box and ask them to donate.

(Student, CA)

Because I had developed a particularly good rapport with these two young men, I pressed these issues a little further. I went on to talk about some of the food activists and farmers that I had met around the Berkeley area, who were interested in healthy and alternative food, and who were not white. I explained to the two boys that in my understanding, being black does not necessarily mean that someone is uninterested in the type of food we produce in their SGCP. I asked these two students, "So, what do you think about that?" They answered:

Student G: I have white friends who don't eat healthy.

(Student, CA)

Student F: I have some white cousins and they eat good food just like the other black people. So like, alright, [but] any healthy food we eat, all you gonna see is like 5 or 10 black people eating like white people food.

(Student, CA)

Student G: This is not racist or nothing, but black people use grease and white people use olive oil, why do they do that . . . because it doesn't have fat in it?

(Student, CA)

To these boys, the association between racial identity and food is very firmly bounded; even if black people do sometimes eat healthy food, they perceive this to be an irregularity in the "normal" patterns of black behavior. In this view, then, even their own practices of eating healthy food within the SGCP would not be enough to shift the boundaries of food-based racial identity. Instead, they would simply be temporarily eating someone else's (white) food.

Teaching race

During the course of my research, I went on to question several SGCP teachers about this frequently articulated distinction between "white food" and "black food." Many teachers had also heard their students articulate these terms, and offered a variety of explanations. In talking about food and racial categories with a black male-identified teacher at Central School in Berkeley, the interview revealed a story about the visceral economy of access that is arguably present within any eating event within the SGCPs. Importantly, it is an economy that goes beyond the food itself:

> *Teacher*: I think, to be honest, I think that what takes [black students] out of their comfort zone, more than the food itself, is how the kitchen operates, and the dynamics behind how the garden operates. It is less the food itself and more the other things connected with the gathering and the readying and the preparing and cooking and partaking of the food, that style is what I think that they truly perceive as more "white."
>
> *JHC*: Because it is not what they are used to?
>
> *Teacher*: I wouldn't say that is not what they are used to, because everyone [prepares food]. But it sounds a little bit different, and smells a little bit different, and feels a little bit different. I mean in any household, division of labor has taken place, and the pans are rattling around the kitchen either way, but the sounds you hear, the smells you smell, the feel, the vibe, and the emotions are just coming from a different place.
>
> (Teacher, CA)

This teacher's description is an important one for what it reveals about embodied food experiences and racial identity. It reminds me of geographer Arun Saldanha's (2005) work on the "viscosities" of race, which similarly hints at how comfort and social identity might be linked through such visceral economies of access. That is, it is not the presence or absence of one particular item (e.g. a type of food) that matters to securing racial identity and a subsequent sense of belonging, but rather it is a whole network of relationships, patterns and ways of behaving that convey a sense of familiarity or alienation. As this teacher's description suggests, the way that things are done in the SGCP classrooms—how cooking is approached, how smells emerge—offers many white students a sense of comfort, while instead students of color often feel strangeness or isolation.

Importantly and obviously, however, certain people also tend to be in *control* of this seemingly diffuse network of relationships and behaviors. And, the people in control of the SGCPs spaces—that is, of the school district, the school, and the classrooms—also tend to be white. Thus, as a black student articulated to me, the food is also "white" not because the SGCPs are individually or culinary racist, but because the broader context in which power takes hold in the class-room is largely a white-centered context: "I don't think that they are racist, like they don't like black people, I just think that since most of the people in the school district are white, and white rules everything, it's gonna be white food" (Student, CA).

Beyond the kitchen classrooms, several SGCP teachers also noted that many of the students of color did not like to work in the garden. This was true of students in general, but the majority of the students of color who I spoke with also mentioned that they particularly did not like to get their shoes or clothes dirty. Some teachers seemed to think that the issue had to do with the historical memory of the black community, particularly in regard to connotations between working in the garden and slavery. A teacher at a related Berkeley school recounted a story to me as I was visiting the school one day:

Field notes, December 2007, CA

She mentioned that at [her school] she had had some conflicts about gardening and racial hierarchy. Like, she said, one kid yelled at her and said "You aren't going to get me to put my hands in that dirt. I am not working for you." She explained to me that gardening seems to have this stereotype for certain people, not only black but also Hispanic kids, that the teachers want to get away from. She said, "So, what do I do with that? How can I tell him that this isn't what it is all about?"

Here we might ask why this child needs to be told that, "this isn't what it is all about?" What if the point, for him, is that it *is* what it is about—that to him it *feels* like oppression? While some SGCP students (and teachers) might have positive (re)actions to being in a garden, perhaps because their past experiences are legitimized and confirmed as "good," we can imagine that other students might not have access (in a broad, visceral sense) to these same positive (re)actions. These students may feel undervalued or misunderstood when teachers try to distance the group from these negative responses, or when they try to explain away such reactions as not really "what it is all about." This eschewing of negativity disallows such students from playing an active role in (re)defining the purpose of their SGCP, and ignores the diversity of identities that SGCPs touch in a variety of different ways.

Another teacher in the Berkeley school system that I interviewed also admitted that there are "definite differences" in how students experienced the program based on class or race. Yet, this teacher did not seem to want to actively address these differences, insisting instead on the power of repetition to *overcome* them:

> I think the concept of getting dirty is one interesting thing. Like, when we first started [a recycling program] the middle-class kids could do it, no big deal, whereas the ones who had relatives who would recycle for a living, it was too close to home. That was lowering yourself. And when they did farm field trips some Latino parents were upset—cultural sensitivities. I think because there has been more access and experience, those experiences aren't as foreign, and there is more openness with regard to food, etc.
>
> (Teacher, CA)

While SGCP teachers often talked about issues of race and class informally with their students, there did not seem to be a common practice of addressing these issues as part of the formal instruction within SGCP classrooms. Perhaps this is due to the fact that there is very little formal instruction in SGCP classrooms overall, leaving such issues to be taken up in other classrooms (in which I did not spend time during my field work). Of course, this division of lived space and intellectual space is not uncommon in schools, but it is nevertheless questionable as a SGCP practice, in a place where hands-on learning takes primacy. While schools teach about race and class abstractly through social studies and history, they often fail to address how social identities are also reproduced in the daily practice of life at school in ways that structure students' tendencies of feeling and behaving, and lead to differences in their motivations to *do*, including their motivations to eat. Ultimately, though, it is this living out of identity in school that comes to matter to students' academic "success"—both in terms of how it is defined and how it is experienced (Beattie 2002).

Undoubtedly the reproduction of race and class hierarchies is a sensitive subject that is not easily discussed in school settings, and that also threatens to undermine the authority of the school itself as a disciplining force (which I would argue is probably ultimately productive). The subject is also certainly sensitive for students, who may not feel comfortable discussing their economic or social struggles and inequities with their peers or teachers. As I will discuss in more depth in Chapter 8, however, SGCPs could help to bridge this gap between students lived, visceral realities and their abstract, intellectual work in ways that would make education overall a more relevant and productive part of their lives. Indeed, from the experiences that I had within Central School's SGCP, food seems like a good starting place for schools to begin to bridge this gap. Yet such work was avoided in Central School's SGCP particularly because it brought up negative feelings —shame, embarrassment, anger, sadness, etc.—that the teachers did not want to encourage or allow. Instead, teachers often encouraged SGCPs to remain a space of what I came to call "positive politics," where the goal was to encourage only positive experiences. There is value in this attempt (Brison 2003), and yet there is also missed opportunity (Probyn 2000), particularly in terms of addressing more directly issues of race and other social differences.

In my interview with the same black male-identified teacher I mention above, we continued our conversation regarding SGCP students' identification with black food and white food. He suggested to me an important connection between students' racial identity and their experiences with poverty. For him, the issue was not ultimately a racial one but instead came down to one of economic privilege. Because many of the black students at Central School were also poorer than many of the white students, they grew up in families that did not have the purchasing power to buy fresher, healthier foods, nor the time to cook them. His explanation illustrates not so much a desire to keep racial identity flexible, as an attempt to eschew race altogether in favor of more straightforward, easily explainable causes:

> What [the black students] don't understand is that those are the food choices that the economy makes you make . . . and now I got my associates degree and my bachelors degree and I started eating healthier, and I feel a lot better. So I am not sure that the students who told you that it was white food, they just don't have the life experience to understand that what they perceive with

a color is actually healthy or less healthy and that that dynamic is directly related to your economic position.

(Teacher, CA)

In this explanation, it is again the students who do not understand, and the (educated) teachers who can offer the students clarification. While this teacher's analysis is valuable (especially in exploring the intersections of race and class), his black students' identification with food in racial terms is also significant and real. Yet when I asked him later if he felt this was something the school should address explicitly in SGCP classes, he said no; "perhaps in a black student union" or something like this, but not during regular school hours. The implication was that school was not a place to try to tackle these potentially difficult matters.

Nevertheless, regardless of how the black students with whom I spoke understand their economic position vis-à-vis other white students at Central School, it is clear that these issues of race and class are present within SGCPs, even as they are not discussed formally. Students of color's experiences of SGCP's cooking and gardening activities (at times) reinforced racially based food identities by confirming their perception of a black food/white food distinction, which is based upon the food's "healthfulness" and understood in fixed terms. This perception was an issue that also concerned all of the nonwhite food activists with whom I spoke while in Berkeley, and one that many such activists are aggressively trying to disrupt through various projects and educational initiatives within and beyond schools (see, for example, the Farm Fresh Choice project of Berkeley's Ecology Center; www. ecologycenter.org/ffc). But, as several of these activists commented, the problem with school-based SGCPs is that the ratio of white to nonwhite teachers in Berkeley is still very high; thus, such associations continue to be made in part because most of those in a position of "authority" in SGCPs are white (and/or middle class, and highly educated). Moreover, beyond the schools themselves, as the students above discussed, the other alternative food spaces in Berkeley—farmers markets, organic food stores—are also coded as white. By not dealing with these issues directly, therefore, the classrooms of SGCPs too become reinforced as white spaces.

Boundary events

To label SGCPs white spaces, however, is just half of the story—the half that, perhaps, leaves agency to the dominant racial group. We need to ask also about how the perceptions and boundaries of race-based food identity are resisted and repracticed within the spaces of SGCP classrooms. And, we need to attend to how the students of color themselves are an active part of this process. Indeed, if black students are in fact a dominant group (percentage wise) within SGCP classrooms, then how could they *not* also be engaged in remaking their SGCP experiences in ways that make them less "white"? Over the course of my research in Berkeley, I witnessed many examples of what I came to call "boundary events" (Twine 1996), which are events in which the boundaries of racialized (as well as classed/ gendered/aged) food identities are called into question. I want to begin with the claim below, grounded simply in the opportunity that SGCPs provide for students to eat together:

> It's promoting fundamental human contact, that cooking and eating is a basic human activity, and we only share food with people who are near and dear to us. Eating is celebratory, it's many things, it's about humanity. And if we are going to talk about a world at peace and a world that respects diversity, then a lot of it is how do we share food. How many black people do white people invite to their homes to eat? How many blacks invite white people? And the cooking program is about that sharing. They eat the food they make collectively. That's a very rare opportunity and for nothing else, forget organics, that's invaluable in itself. But we have so much strife in our world, this is an opportunity for peace and love and sharing.
>
> (Leader, CA)

Here we can again see, as in previous chapters, that the work of SGCPs is also about providing opportunity. Regardless of how the food is coded—as black or white, healthy or unhealthy, delicious or nasty—SGCPs do provide students with unique opportunities to relate to each other through food. While there are undoubtedly hierarchical forces at play within the context of the school, the outcome of this relating within SGCPs is not predetermined by such forces. In fact, the opportunity of relating opens up the possibility for new experiences and perceptions of food to emerge.

Furthermore, as food scholar Rachel Slocum's work (2007, 2008) makes clear, the racial coding of certain foods and food spaces is not always or only a negative. As she explains of the farmers market customers in her study, "far from simply producing distance and cordoned off spaces, whiteness is also a process of reaching out toward brownness via efforts, in this case, to bring good food to others" (2007, 7). Here she is talking about the desire of white alternative food activists to reach out to nonwhite people, and to change the "face" of whiteness (a desire I witnessed often in this SGCP research). As Slocum explains, a bodily desire for proximity to brownness can also be a part of the "tendencies" of whiteness, and can therefore also lead to new outcomes and connections. Certainly this is part of the story of SGCPs, where a diverse group of students are put into contact on a daily basis with a variety of (un)familiar foods and food spaces— a connection that in and of itself permits a certain sort of access. Along these lines, Slocum, quoting Rajchman, suggests that to connect, therefore, is "to work with other possibilities, not already given" (2007, 7). In other words, in connecting we might find that tendencies can change, and so too our bodily habits of feeling.

Re-teaching race

One of the cooking teachers in Central School, Ms. Tiffany, was a young, educated, middle-class African American woman who was extremely knowledgeable about alternative, healthy eating. Her presence in the cooking room frequently called into question the boundedness of healthy food as "white food," even as the students of color insisted on this boundedness. Ms. Tiffany frequently shared her interests in holistic nutrition with the SGCP students, and sometimes she did so in ways that created tension among black students who were uncomfortable with her (lack of) performance of a (typical) black identity. However, I came to recognize that these moments of tension were also productive moments, where the act of relating to Ms. Tiffany allowed students to recognize that not all "black" bodies prefer the same kinds of food. She explained her role in the classroom in the following way in regard to black students:

> Before I am African American, I am human, and I always try to interact with kids on a very basic human level. But I think that [it] would be really good, I suppose, [that] I would be a good role model for the African American kids in particular, probably, that it is nice for them to see a different representation; do you know what I mean?
>
> (Teacher, CA)

Another teacher similarly reflected on the visibility of teachers in the SGCP classrooms: "When we give people role models that look like them, they tacitly understand that they could be them. It has nothing to do with food" (Teacher, CA).

And yet, in the context of SGCPs, such visible identity markers also have everything to do with food. In relating to Ms. Tiffany as simultaneously a black woman and also someone who was deeply interested in healthy alternative food, students' rigid perceptions of race-based food identities could be called into question. In my field journal, I commented on Ms. Tiffany's relationship with her students:

Field notes, November 2007, CA

Pete and Derron were talking to Ms. Tiffany about racism, and Pete was saying that his grandma was hecka racist because she would always say mean things about the white kids he was skateboarding with. Ms. Tiffany explained that his grandma lived through some tough times in American history, and that she was probably reacting to this. Later that day, one of the boys was climbing on a high wall, and Ms. Tiffany told him to get down. Derron said, you told him to get down cause he is black right? She laughed, and said yes, right (sarcastically). He obviously had said it to get a reaction. I realized that this SGCP space was a sort of open space in which to test out these ideas, to joke around. Like, some of the black girls were teaching Tim and Matt [two white boys] how to dance to hip-hop, and everyone was laughing and getting into it.

Later, Ms. Tiffany explained to me:

> There are certain ways that I relate to [the black students], because there are
> certain things that I understand, in coming from a black family and a black
> culture, like having certain relatives that are old school and will say certain
> things, or how someone's granny [cooks, selects ingredients], so I can get it,
> or understand. But . . . there are some things that I don't relate to them on
> . . . Like just the other day, Benny was asking us about different rappers, and
> it is just like sometimes, they might feel like, maybe I am not black because
> I don't know every rapper. It is a tough and loaded question for me, because
> I probably come, education wise, and growing up in a suburb, and going to
> really good schools and stuff, I probably come from more of a background
> like yours [white middle-class researcher] than theirs . . . So, like Leah for
> example, I know she really likes me. But there are also the girls that don't
> like me, that cut their eyes at me and maybe feel like, she is black, but she is
> not black. And somehow are annoyed by that.
>
> (Teacher, CA)

In this case, Tiffany's ability to relate to both "black" and "white" sets of associations
put her in a precarious position vis-à-vis some of the black students in her class.
Some black students became visibly angry or confused at Tiffany's behavior when
it did not conform to their understandings of "blackness." Perhaps also, students
could read her conformity to "white" foods as a sort of practice of assimilation
into "white" culture, which is threatening to the continuation of "blackness," and
thus uncomfortable. However, and despite these tendencies for race to congeal,
such moments of tension and *dis*comfort also illustrate that the tendencies of race
(or gender, class, etc.) are far from fixed.

There were many moments of re-evaluation and re-interpretation over the course
of my SGCP research, particularly in regard to issues of race. My initial impressions
of the Berkeley SGCP, for example, included feeling put off by the fact that one
of the cooking teachers (a middle-class white woman) began each class with the
gong of a singing/meditation bowl—which is also a very typical way to begin and
end a yoga session, a common practice among wealthier Berkeley residents. Because
of my interviews with food activists of color, who called into question the ability
for white teachers to teach nonwhite students about food, this bowl-ringing
practice initially struck me as elitist, and potentially off-putting to students for whom
this practice is unfamiliar or weird (at the same time as could be legitimizing and
animating for those who are familiar with meditation). In other words, within the
context of Berkeley, this singing bowl represented to me the side of Berkeley that
students described as hippie/yuppie and white, and thus I interpreted it as a potential
barrier of access for students from poor and nonwhite households.

Yet, as I came to recognize, few students seemed to mind the sound of the
bowl, and many were actually interested in ringing it themselves. The bowl had
a gentle tone that felt soothing to me, even as I interpreted it as elite. When I
brought up my initial reaction to one teacher during an interview, she responded:

I suppose that the program could be more sensitive in terms of [recognizing differences among students' comfortableness with the space/food] but it is also just about who is running the classes. [The cooking teacher] brings her own sensibilities, and another teacher would bring theirs. Like, she starts the class with a singing bowl, and although not all kids are familiar with that sound, it is a nice sound and a nice tradition.

(Teacher, CA)

While I think that my initial juxtaposing of the meditation bowl experience to the concerns of food activists of color was a valid response, as this teacher points out, this was certainly not the only thing that the meditation bowl allowed. I agree with the concerns voiced by food activists of color that suggest that we need to ask, "What can these institutions do so they can be empowering to a wider range of students?" (Activist, CA). But, I also want to suggest that even in being familiar to some, and unfamiliar to others, the meditation bowl *can be* empowering to both. To not recognize this, I think, is to "discredit students' ability to learn" (Activist, CA) from their interactions with a wide range of people and experiences, and to use this knowledge in creative and significant ways. In this sense, it is important to point out that many students (both white and of color, rich and poor) enjoyed taking turns ringing the meditation bowl.

Playing with race

Though race was not formally discussed in relation to food within SGCP classrooms, neither was it the "elephant" in the room. Students took up the topic of race frequently in their interactions with each other, testing boundaries and "playing" with race as a marker of identity. In fact, some students even suggested that this was part of what occurred during my interviews with students. When I mentioned to another group of black students (all female-identified) that in a previous interview some students had called the SGCP food "white food," their response suggested a sense of play:

JHC: [Kids] I was talking to earlier called the food we eat here white food. [students laugh]
Student H: They were trying to be funny.

(Student, CA)

[students all talking at once]
Student J: It's not white food, really, there is no such thing as white food or black food; it is like saying fried chicken is black food. They were playin'.

(Student, CA)

JHC: So you guys don't think it is true?
Student H: Well, yeah but . . . [it's more complicated than that].

(Students, CA)

Through interacting with SGCP students in the cooking and gardening classrooms, I came to realize that unstructured "play" was an important part of their experience of the SGCP, and an important way in which they were able to make the SGCP experience "their own." In part, this play was encouraged in SGCPs by the great amount of flexibility that these classrooms offered, in relation to the students other classroom experiences. That is to say, the lived experiences of SGCP classrooms were less controlled and routinized than in other more instructionally based classes (an issue I discuss in more depth in Chapter 8). This was especially true in the kitchen classroom, but also somewhat in the garden. Students were at liberty to turn on the radio, sing or dance while cooking, tell stories or discuss life problems while chopping, tell jokes, etc. There were also more formal ways in which play was "structured in" to the classroom experience: i.e. the cooking contest days, or a rainy-day game called garden jeopardy, or the wooden-spoon activity discussed below:

Field notes, November 2007, CA

We had to wait for the potatoes to cook for a while, so while we waited, Ms. Lisa decided to play a game in which students were blindfolded and had a big wooden spoon in their hand. The blindfolded student had another student placed in from of them, and they had to figure out who the student was by touching them with this big spoon. The game was very interesting, because race and gender became quickly obvious as two important factors in determining who someone was. If the student had long hair or short hair, in a ponytail, or braids, etc., they were more readily identifiable. The kids started joking around and giving false clues like, it is a black girl, when it was a white boy, and stuff like that. They made comments about someone's braids being fake, when actually the student was white and did not have braids. Everyone was laughing.

Although these games did not directly address issues of social identity and food, they reproduced the SGCP cooking classroom as a space in which these issues were "in play," and therefore not fixed or predetermined. In this sense, the cooking classroom became a space where all students were actively engaged in the creative production of food knowledge, and the creative negotiation of food identity. During an interview with Ms. Tiffany, I asked her how she would reply to the critique that alternative food spaces like SGCPs are "white" spaces. She responded by illustrating how important students' agency is to the daily running of the cooking classroom:

I am also finding that some kids who come from disadvantaged neighborhoods, they really take pride in what they are creating here, and also that they are capable of that. Especially at [my table], some of the girls that won the last cooking competition, they just went all out, got really into it. They were like, no don't put that tablecloth on there, put this one. They switched the one I had put out for them. You know, so they really take ownership over how the table looks. So, that is really positive ... and so, you know, I think what is racist is actually saying that these programs are "white," because it [discounts the work that] these kids put into it.

(Teacher, CA)

As Ms. Tiffany relays, in analyzing SGCPs, it is crucial to account for how students take up the task of healthy eating and make it "their own." At the same time that SGCPs can encourage healthy alternative eating that is typically coded as white, they also can provide the opportunity for students to actively relate to such food in ways that call into question this racial code. In addition to students' agency, however, if we label SGCPs as simply white spaces, we also discount the ability of teachers (both white and nonwhite) to disrupt this label by inspiring students in a variety of ways to make the food "their own." Indeed, the strength and depth of the relationships between teachers and students, also can allow for a disruption of the boundaries between black and white. As one black girl said to me during an interview, "I consider you guys [to be] role models, because you are actually doing a program that makes us change the way we like food."

What's more, although the alternative/nutrition knowledge that such SGCPs impart is undoubtedly a *partial* knowledge, and is often associated with white culture and identity, it can nevertheless be an important tool for students' lives—including especially those students who do not encounter such information elsewhere. As a food activist in Berkeley said of school and community gardens:

Talking about empowerment of the black community, it is something we can have control over. We often feel like we can't control much, but we can take this into our own hands. We can grow our own food. It is very empowering in that way.

(Activist, CA)

Similarly, some of my student interviewees, and especially those from poorer families and/or mixed-race families and families of color, suggested that the information that their SGCP imparted was important to them because it gave them new ideas for how to respond to various life contexts and experiences:

Yeah, I bring some of the recipes home ... because I cook at home a lot. My mom works, so I can make a healthy dinner for her and my sisters and my grandma.

(Student, CA)

It teaches us new things, like if I was at home, my mom would have never thought about cooking pears.

(Student, CA)

> If we learn how to garden, we can eat the stuff that we do in class at home, and we can have the ingredients.
>
> (Student, NS)

SGCP leaders and teachers also often looked at the SGCP in this way, seeing their role as one of providing important information and ideas that could be useful in broadening the life options particularly of disenfranchised students. These sentiments bring me back to the discussion that I began in Chapter 3, regarding the importance of seeing SGCPs as providing opportunity, even at the same time as they produce limitations and restrictions. As the information (and food) that the SGCP provides helps to disrupt the structures that limit certain students' access (in economic, geographic, cultural and visceral terms) to both certain foods and food ideas, it also simultaneously helps to "unstick" these foods and ideas from potentially unhelpful patterns of racial and ethnic coding. By doing so, such SGCPs encourage students to expand their relational reach, allowing them increased opportunities to find agency and produce meaning within an undoubtedly uneven and unjust food system.

Concluding thoughts

> We need to realize that cooking and gardening issues are everyone's issues; gender or race is not important.
>
> (Teacher, CA)

The above passage from a teacher at Central School raises important questions about the nature of food-based knowledge within SGCPs, and about how SGCP leaders imagine this knowledge to be connected, or disconnected, from social identity. Given the examples from this chapter, we might ask, if cooking and gardening issues are important to everyone, then in what ways could gender or race *not* be important? By examining the particularities of food and racial identity, this chapter has provided new insights into the idea that cooking and gardening issues are "everyone's issues." While the examples discussed above show that students experienced the SGCPs in a variety of ways, they also illustrate that the SGCP classrooms were relevant for many if not most students, even if they are incongruently so. Although this does not mean that SGCP classrooms are equally important to all students, it does mean that there are opportunities for recognizing and highlighting racial diversity in ways that encourage racial "playing" and disrupt unhelpful stereotypes.

However, lest we get too caught up in the romantic, creative potential of relational thinking, a warning also comes along with these analyses. The erecting and dismantling of borders and boundaries, whether real or imagined, is itself fraught with uneven power relations. The experimentation and play that is called for in such conscious, embodied openness rarely comes in our society without heavy prices: public disenfranchisement, violence, poverty, and physical confinement (McWhorter 1999). Indeed, we rarely experience the type of freedom that would allow for such unhindered visceral experimentation or play. In this sense, and as the above experiences confirm, "the work of anti-normalizing ethical self-transformation cannot be fully separated in practice from the work of cultural and institutional challenge and dismantling" (McWhorter 1999, 225). "Counterattacks" must happen at multiple scales simultaneously—locally, nationally, and globally. We must ensure that we provide each other with the spaces and tools that are required to play, experiment, and grow.

6   Real men eat raw onions

I don't think [gender makes a difference]. There is as much enthusiasm from the boys as the girls, and the fathers will come to the [cooking events] too.

(Teacher, NS)

Well, at home, with some parents, the mom mostly still cooks, but cooking is more often shared now. In our house, cooking is shared.

(Mother, CA)

Feminist scholars and activists have noted the invisibility of gender in everyday life, lamenting the (mistaken) perception that gender struggles are a thing of the past (Valenti 2007). Gender seems to remain conspicuously hidden, in plain sight—hidden behind legitimate experiences of gender equity, as well as continued assumptions about the normalcy or the naturalness of gendered difference (Serano 2009). The idea that race or class difference matters to people's food experiences is something that food activists and enthusiasts have surely begun to recognize. But gender differences seem to be generally under-considered (though see Orenstein 2010) perhaps because gender hierarchy itself is presumed to be a historic matter, and not a contemporary one. After all, some of my interviewees reasoned, both boys and girls are taking garden and cooking classes now.

For better or worse, history lingers in the present more than we might like (Kobayashi 1994), and as a result, embedded gender norms and expectations come to matter within SGCPs in many different, though perhaps elusive ways. During my interviews with teachers and parents especially, many of my direct gender-focused questions were largely dismissed as no longer relevant, mainly for the reasons expressed above. Cooking at home is thought to be more evenly distributed now, according to my interviewees, or at least it is shared evenly among progressive, gender-conforming, heterosexual couples living in two-parent homes. These perceptions, however true on a family-by-family basis, stand in stark contrast to broader contemporary statistics on household labor—statistics that tell us that little about the household division of labor in North America has actually changed in 20 years (Hochschild and Machung 2012).

My research experiences overall confirm the assessment of many feminist food scholars that gender impacts the formation of food-based identity in important though perhaps under-recognized ways (Avakian and Haber 2005; Williams-

Forson and Counihan 2013). Many of my adult interviewees sensed that the gender norms surrounding food and cooking have changed with the current generation of parents, such that both self-identified boys and girls would feel equally comfortable in garden and cooking classrooms. And indeed this is true; many of the students that I came to know through the cooking and gardening classes told me that both women and men could cook, and that gender therefore didn't matter anymore, at least in the kitchen. This sentiment does not explain, however, the disproportionate number of self-identified women in every aspect of the SGCPs that I researched—in leading, teaching and volunteering. It does not explain why most of the parents who answered my requests for interviews about food and cooking were actually mothers (in heterosexual, two-parent homes). And, it also does not explain the differences in both behavior and valuation of the programs that I witnessed between male-identified and female-identified SGCP students. Certainly gender has an influence on how students—and adults—experience these programs, even if we all agree that the kitchen (and garden) should be a gender-inclusive space. In the following sections, I first explore how the SGCPs that I researched were involved in the reproduction and enforcement of traditional gender roles, particularly in regard to the practices of cooking and eating. I then go on to explore how the SGCPs also became spaces in which these gendered norms and practices could be tested and disrupted, especially by the students themselves.

A few notes about gender and language are important here. In this chapter, I use the terms "male-identified" and "female-identified" (or variations thereof) to highlight that it is the preferred gender identity of the students (and adults), rather than the gender assigned to them at birth, that matters most to how we imagine gender to operate in the spaces of the SGCP. Wherever possible, I also come back to the language of gender diversity more broadly in order to signal that gender identities do exist that do not conform to the binary of boy/girl that all of my interviewees employ in their discussions of gender difference. Moreover, although experiences of transgender and gender nonconforming students were not directly reflected in my case studies, this does not mean that these experiences do not exist. For example, trans women can experience eating disorders at higher rates than cisgender women (Vocks et al. 2009). Overall, what I hope to illustrate through the empirical examples in this chapter is that gender as a social category is more changing and changeable (for all persons) than the binary of male/female allows, and also that SGCPs can work to allow and encourage this changeability by becoming cognizant of the ways that traditional gender norms operate, and are challenged, within the practices of the garden and kitchen classrooms.

Food as feminine

> There is something motherly and nurturing about gardening and cooking. The first time I worked in a garden and we were harvesting and we had armfuls of squash and it was like carrying babies. It's nurturing. Very feminine. And for me in my house my mom did all the cooking.
>
> (Teacher, CA)

Some women are able to connect to the nurturing and emotional side of these programs, and I think people who are food activists, I mean, there are a lot of men involved, but there are [more women].

(Teacher, CA)

The most striking way that gender difference is present within this research is the fact that the vast majority of my interviewees were women—from the SGCP leaders, to dietitians, to teachers and to mothers. This amounts to 85 percent of a total of 100 interviews with adults. More than any interview quotation or field note excerpt, this fact alone suggests that gender still does matter a great deal when talking about, and engaging in, food-related issues. The realm of food, at least as a matter linked to the feeding of children, is still largely "women's work" within (and beyond) SGCPs, regardless of whether or not it is always explicitly viewed this way. I questioned many of my interviewees about why it was that so many women were involved in every aspect of SGCPs, and their answers were various, including cultural ideas about the perceived femininity that is embedded in the act of harvesting produce or cooking (as in the quotation above) to more structural issues regarding salary and unpaid labor. As two volunteers in the Berkeley program explained of the latter issue:

I wish there were more men involved. For women we, there is a layer of feeling like [we] can deal with less money and men need to provide more money . . . Yeah it bothers me, it shouldn't be all women. I mean we like this work but we want to be doing the policy decision- making too. It would be interesting to do the gender breakdown.

(Teacher, CA)

One interesting thing I started to feel about [a prominent volunteer organiza- tion] is that comparing women's job searches and men's, I think it's hard for men to accept the amount of money we get because they are feeling a pressure to earn money. I know [another guy] was [in the volunteer organization] last year, but he has had problems with the fact his girlfriend makes more money than he does, so that is a big deal getting into this field.

(Teacher, CA)

It is well known that women have taken on, and continue to take on, signifi- cantly more unpaid care work than men over the course of their lives (England 2003, Miranda 2011, Hochschild and Machung 2012). What constitutes "volun- teerism" versus other forms of care work is a tricky question (Taniguchi 2006), often based on more formalized opportunities to engage in care work. The two female-identified teachers above were both involved with SGCPs through a prominent volunteer organization that offers small stipends in exchange for full- time work within the SGCPs. As their comments suggest, these women are aware that it is more socially acceptable for young women to make less money than it is for young men, and they lament that this means that the burden of their food-based care work falls disproportionately on their shoulders. It is clear that these women believe in the work that they do—that they deem it to be of great social importance—and yet they simultaneously wish the work to be shared more evenly, so that they might have opportunities to be involved in other aspects of the food system, in places where they might gain more power, and possibly make more money. It is also worth pointing out that, in other contexts, women activists have purposely embraced the cultural association between femininity and food-based care work in order to gain more political power and achieve their broader social goals through food activism (Counihan 2008, Allen and Sachs 2013). In this sense, while not providing financial gain, food-based care work has afforded some women opportunities for political power that, if not couched in terms of traditional household labor roles, might have come to be challenged. Thus, we could also read women's involvement in food-based care work with some amount of ambivalence on the part of women activists, whereby certain seemingly oppressive gender roles are embraced in order to achieve push back elsewhere in our social system.

Whatever the reasons are for these women to feel compelled to engage in this food-based care work, the fact that more women than men occupy a variety of roles within SGCPs is also important to how the programs are experienced by those who take part in them. Considering that women disproportionately occupy positions, including leadership roles, within the SGCPs (that is, as instructors, or ideologues or adult volunteers), we might ask what this proportionality teaches kids about who belongs in the cooking and gardening spaces, and who cares most about this work. If we agree that "when we give people role models that look like them, they tacitly understand that they could be them" (Teacher, CA), then how could gender *not* come to matter under these circumstances? As one mother commented to me:

> We do learn in the subconscious . . . I think [the person who teaches the kids] could have more of an impact on who cooks at home. I think the more male teachers you have in general for the [SGCP] classes the better. Balance the [presence of] female teachers.
>
> (Mother, CA)

Besides the question of who is doing the teaching, there is also the issue of how "women's work" within the SGCPs is compared to or valued against "men's work" outside of the program. Undoubtedly the reduced salaries (and heavy volunteer workloads) of the two women above suggest that this work is monetarily undervalued in our society. This is especially true considering that these women's salaries/stipends do nothing to reflect the great amount of interest that so many in North America have expressed in regard to issues of healthy eating, obesity, and dieting—and especially in regard to how we feed our children. Of course, this is an issue within the teaching profession in general, and is therefore not just related to SGCPs. Nevertheless, it is telling that in my own research, some of the harshest push back that I heard regarding the compulsory cooking and gardening classes came from several traditionally masculine fathers, who questioned the value of learning to cook in comparison to courses in math or science, for example. The sentiment was clearly that cooking courses were an extra, unnecessary and unhelpful in helping students to get ahead and work towards a career, where as math and science might actually make them socially successful.

Given these sentiments, it is important to consider what the lack of male-identified role models in SGCPs might teach students about the social worth of a cooking teacher or a cafeteria worker, as opposed to, for example, a professional chef. As I mentioned briefly in Chapter 4, The Nova Scotia case study site gave me an opportunity to explore this issue with a number of my interviewees. Below I discuss with a SGCP leader the juxtaposition of the "lunch lady" figure, Ms. Dora, who teaches the children how to cook on a daily basis, and the professional chef, who visited the school once a year (and, as I detailed in Chapter 4, created tension with some of the parents).

> *Leader*: Oh I'm sure they love Ms. Dora, but when you got the chef, who is a professional chef, wearing his uniform, I think that they take something more from that. I mean working with Ms. Dora is great for them and they love doing it.
>
> *JHC*: Did he wear a hat?
>
> *Leader*: Yeah. He's all decked out.
>
> *JHC*: So it was obvious that he was a celebrity chef?
>
> *Leader*: Oh yeah, because he comes in and just the way he cuts the things. He doesn't even look half the time [makes swishing sounds for cutting] and he's cutting this and putting this in here and you can just tell by the way he moves that this is what he does *professionally* and the kids pick up on that. It's kinda like, you can play baseball with me, or you can play baseball with a professional baseball player. You know, I'm great to play with but . . . you know, what are you going to remember when [you're older]? . . . it's that kind of thing.
>
> (Leader, NS)

The SGCP leader who is speaking here is a male-identified school administrator, who has little direct involvement with the daily running of the SGCP itself. His interpretation of the importance of the chef versus Ms. Dora, however, is telling of the social worth that is placed on cooking as a "professional" occupation, as opposed to something that is presumably practiced as a domestic task or within a care-giving role. Moreover, his interpretation also underestimates Ms. Dora's own professional skill, considering that Ms. Dora is called a "saint" by many of the other teachers in Plainville School for her ability to teach the kids and keep calm in the kitchen in the midst of much inevitable chaos. It is also important to remember that Ms. Dora's professional work additionally includes feeding lunch to approximately one third of the school on a daily basis, including through the incorporation of the school garden's vegetables. The notion that kids would invariably find the chef event more exciting—and more memorable, even years later—is also curious considering that it is in the small-group interactions with Ms. Dora that students were most able to take control of the kitchen space and make decisions on their own. The chef event was certainly an experience to remember, especially because it was so out of the ordinary, but it was through the seemingly mundane and repetitious tasks of kitchen and garden duty with Ms. Dora that most students actually learned to cook.

The difference in how cooking work is valued, in the above example, has to do with what the work is *for*. In the case of the celebrity chef, it is clear that his cooking work has been elevated from care work or working-class labor to something that has afforded him both power and money. Cooking has made him "successful," and as such, he stands with his tall white chef's hat as a role model for what the kids could achieve. Ms. Dora, on the other hand, does not represent success. Her job as a cafeteria cook has not afforded her great financial stability, nor social fame. She cooks for children as she teaches them how to cook for themselves—a type of work that is not significantly divergent from domestic care work, and as her hairnet symbolizes, is also not quite distinguishable from working-class, "low-skill" labor. To some, then, her kind of work is seen as simply not as valuable, even though it is vitally important to the functioning of most households. This valuation, arguably, mirrors the above discussion of volunteerism by female-identified garden and cooking teachers in California. In both cases, it is important to note that what is most undervalued is also what is typically considered women's work.

I heard this type of undervaluing repeated not just by adults but also by students in both SGCPs as well. Interestingly, often the way that students came to value the cooking and gardening classes was related to how they imagined themselves in food-based roles in the future. Male-identified students frequently justified their presence in these programs by claiming that cooking skills would afford them money and prestige in the years to come. Female-identified students, on the other hand, seemed more apt to reference how learning to cook or garden might help them to fulfill more domestic duties. Comments from male-identified students frequently included assertions like, "I've got skills to pay the bills!" (Field notes, November 2007, CA) a statement that referenced the student's own chopping ability as being marketable. In contrast, when referencing why the programs were valuable to them, female-identified students more frequently referred to present or future instances of domestic labor—for example, "My grandma would put ham in this," or "I am gonna take this [recipe] home so I can make it for my sister and mom" (Field notes, November 2007, CA).

Considering this reproduction of a gender-based division of labor, then, we might ask whether female-identified and male-identified students tend to experience their cooking and gardening classes as relevant to their lives *in different ways*, even as they both express a comfortability in being in the classes in general. That is, while both boys and girls frequently articulated that they liked being in the SGCP classes, and felt equally comfortable in them, their enjoyment and valuation of these programs seemed to come from different(ly gendered) social places. Some of the SGCP cooking teachers that I spoke with agreed with this assessment, offering evidence from the ways that the kitchen duties tended to be divided among students. In their experience, boys were less likely to follow through with "chore-like" activities of washing dishes, or sweeping the floor. Instead, they seemed to be more engaged in traditionally masculine tasks like lifting heavy objects, carrying loads and using sharp knifes. Boys would often get excited about these kitchen duties, and even seek them out, while table cleaning, dish washing, and sweeping were typically avoided (and left to the girls, or their teachers).

In the garden, SGCP garden teachers similarly noted that while both boys and girls complained about getting dirty, the boys would sometimes respond well to tasks that are typically associated with masculinity—lifting heavy objects, digging large holes or doing other physically strenuous work. For example, one male-identified garden teacher that I spoke with even suggested that he could use "the sort of fear factor culture" of valuing strength over weakness to convince male-identified students to taste vegetables that they wouldn't otherwise try: "Like, if I say the radishes are really bitter, so I don't know if those boys will really like it . . . then they [will] eat it, try it" (Teacher, CA).

Such opportunities for (seemingly) daring, risk-taking behavior gave some male-identified students a chance to perform conventional masculinity in otherwise feminine-coded food spaces. The fact that these appeals to masculinity "worked" from the teachers' perspective (e.g. by convincing the male-identified students to try new, healthy foods, or to help out in the kitchen and garden spaces) speaks to the often implicit and yet ubiquitous ways that gender roles are present within popular culture, even or especially among young students (Gilligan 2003). Indeed, it is this ubiquity that ensures that these gender roles are present and understood among students within the context of SGCPs, regardless of the intentions or perceptions of teachers and parents. Male-identified and female-identified students (as well as gender nonconforming students, to be sure) therefore likely experience these programs differently on the basis of the opportunities that these programs afford them to play out, and play upon, their expected gender roles.

Furthermore, differently gendered students also experience these programs differently on the basis of what they bring to the SGCPs from their home life situations as well. Despite what some parents articulated to me regarding shared household chores, many of the students with whom I spoke told me that their moms do most of the food preparation for their households, with dads (if present at all) perhaps cooking on the weekend or on special occasions, involving grilling or meat carving or more typically masculine skills. In addition, a lot of the female-identified students with whom I spoke suggested that they too were involved in their household's daily cooking activities, while male-identified students more frequently expressed that they cooked for their friends or themselves, or perhaps not at all. Given these home life differences, it seems that what male-identified and female-identified students, as well as gender nonconforming students are able to get out of their SGCPs might be dependent on what their home experiences and roles teach them about what is valuable or exciting, as well as where they belong. As one teacher told me:

I also think the level of empowerment I was talking about is more just like feeling like you are doing something positive in your everyday life, like if you learn how to cook something then you can make something. One of the female students, she cooks a lot at her house and she and another woman, you can tell they want to be adult. And, feeling like they can cook at home is really important. And, doing something good when they are doing bad in other classes is really important [for] personal development, whether you argue it's related to food or not. Even just feeling like you have a role in creating something. That's not standard for many kids on multiple class levels.

(Teacher, CA)

In this teacher's description, the female-identified student's ability to get something out of the SGCP is based at least partly on her cooking role at home. It is because she cooks for her household, and because they depend on her for this, that she is able to feel particularly empowered by the cooking classes, despite not excelling in other areas of her schoolwork. While arguably both male-identified and female-identified students could experience this kind of empowerment, the broader social expectations surrounding gendered food identities may delimit the circumstances under which different students tend to embody this sense of empowerment.

Gendered eating practices

In addition to the practice of preparing food in general, the *particular* foods that students prepare in SGCPs are also gendered in significant ways. Most significantly, "healthy" foods, which carry connotations of fewer calories and lesser fat, are often associated with dieting and body image. Although these are issues that have impact across gender identities, they tend to impact women disproportionately (Bordo 1993, Thompson 1994). In one group interview with two male-identified and three female-identified students, I asked the group about whether the program was equally relevant for boys and girls. One of the boys responded, "Wasn't it women who started the idea of dieting?" (Student, CA). He was not alone in thinking this; among many of my student interviewees, it was understood that concerns about fatness and dieting were particularly feminine concerns, or at least were "female" in origin, even if boys also engaged in conversations about obesity. In so much as SGCPs were perceived to be gendered (as feminine), the programs were also thought to be focused on the specific goal of preventing bodily fatness, through both nutrition education and healthy cooking.

In my participant observation in SGCP classrooms, I witnessed both girls and boys talking about the fat content in various foods. However, the word *fattening*, a term that particularly relates to body size, was a word that I heard used particularly often by female-identified students, along with the phrase "pig out." In day-to-day conversations about food within the SGCP classroom spaces, it was the female-identified students that vocally wondered, and worried, about fatness. The following experience was not uncommon during my participant observations:

Field notes, November 2007, CA

I was sitting at table three, helping Latisha, Shaunna, and Paula make tortillas, while two other students were sautéing veggies. The girls were talking about what they put in various dishes that they make at home, and Latisha mentioned bacon in a rice dish. Shaunna said that bacon had fat in it, and was *fattening*. They started talking about cheese, and asking each other if cheese was bad. Paula said, "Yeah, cheese has like lard in it, or like fat from an animal or something, so it is bad. *It will make you fat.*" The others agreed (emphasis added).

The adults within the SGCPs were well aware of their students' conversations about dieting, and were cognizant of the girls' worries about fatness and body size. When I mentioned issues of negative body image, or the potential development of restrictive eating habits (e.g. anorexia or bulimia), many of my adult research participants nodded their heads knowingly and expressed concern over the issue. "There are so many pressures on students to diet . . . especially girls" (Teacher, CA). Nevertheless, many of my research participants did not seem to think that SGCPs were a place where eating disorders could develop. Some SGCP teachers and leaders idealistically suggested that students at this age were "still too young" to really be thinking about such issues (Field notes, July 2007, NS), while others simply did not see SGCPs as places where food restricting was encouraged. Yet the students that I talked to expressed otherwise. Dieting came up frequently in my conversations with students, and most often when I was talking to girls. When I asked students whether they found it difficult to eat healthily, female-identified students often responded with a diet-related answer:

> Yeah, I try to not pig out on candy and cookies. It is [hard].
>
> (Student, NS)

> Sometimes I am doing really well, and then I have something sugary and I feel all bloated and huge, and I have to start all over again.
>
> (Student, NS)

> A lot of girls these days are dieting, low-carbing it or whatever . . . I am not on a diet.
>
> (Student, CA)

Regardless of whether SGCPs are consciously framed as "dieting" programs, then, the female-identified students that I talked with often seem to experience them as such. Through being labeled as healthy eating programs, and through encouraging less meat consumption, the SGCPs seemed to give students the impression that they are just another mechanism in the general societal push towards dieting and staying thin. In other words, the students experienced their SGCP as an opportunity to practice the food and dieting lessons that they learn in the broader social world. But sometimes this situation can be frustrating, as one female-identified student described: "We get it at school, we get it on TV; it is everywhere. Sometimes I just want to eat, and not have to think about it" (Student, CA).

At the same time that such food lessons are not gender neutral, neither are they race or class neutral. As feminist researchers have shown, body image and standards of beauty are not experienced in the same ways across lines of social difference (Thompson 1994). In Berkeley, because the SGCPs had connotations of middle-class whiteness, dieting and keeping a slim appearance were also frequently associated with being white, and also with being from a wealthier family. In fact, students of color frequently commented that eating healthily was harder and more expensive than eating unhealthily, despite some leaders' claims to the contrary. A SGCP teacher thoughtfully commented to me:

We don't want to sound like we are preaching or we are better, and to be honest, how are we [better]? Many of us are thin, white women, and you know, I know what a farm market is. But if I didn't, and I went in there and they charged me three dollars for a little pepper, I'd be put off.

(Teacher, CA)

As this teacher points out, motivation to eat in the ways that the SGCPs encourage implies not only a particular choice of foods (healthy and low fat), but also a particular social identity (thin, white, and foodie-centered). Indeed, food choice and social identity are interdependent and co-constituting, a point that previous chapters have already explored in depth. Thus, we might ask, what type of bodily "becoming" do SGCPs encourage when they are assumed to be focused on the task of dieting or the maintenance of a thin body? Do SGCPs motivate students to become a particular type of person, or a particular type of body, above others? And how is this body type gendered, raced, or classed? Another way of asking this is: does the bodily motivation that SGCPs generate limit or enable diversity in terms of food–body identities? One food activist in Berkeley wondered with me: "Are [students] being guided and led by people that are culturally competent enough to create a curriculum that is relevant to them and their real lives?" (Activist, CA). In the example below, a black, female-identified student was in a conversation with several other students at her table, commenting on a dieting plan that she had created for herself for that evening. Her understanding of what constitutes a desirable body size and food practice are clearly linked with particular identities that she does not consider her own. Rather, they are identities that she has learned to value—or at least to identify as valuable—during her experiences within SGCPs classrooms:

Field notes, October 2007, CA

Paula said she would have a salad for dinner tonight, because she was on a diet. I said, "Really, you are on a diet?" And she said, "Yeah, but it is hard because I like salad with cheese and ham and onions and croutons and dressing all on there." She said her mom told her that if she wanted to lose weight, she better eat one of those vegetarian salads with no taste. She and her friends explained to me that the food they eat in the classroom is all vegetarian, like Berkeley hippie food, but the food they eat at home is different.

From the conversation recounted above, we can see that dieting is not only gendered but is experienced intersectionally, as at once a gendered, raced, and also perhaps classed-based phenomenon. For this female-identified student of color particularly, navigating the world of dieting within the SGCP involves decisions that impact her embodied relationship to the culinary attachments and financial status of her home. However trivial or minute such day-to-day dieting decisions may seem, they can have lasting impacts on the kinds of foods, behaviors and body shapes that students see as valuable or desirable. In this student's assessment, her SGCP offers students like her the opportunity to learn how a particular sub-set of Berkeley residents—presumably white and middle-class vegetarians, as the phrase "Berkeley hippie" typically connotes—remains thin. As Paula laments, the problem with these food lessons is that they aren't particularly good tasting, or more exactly, they don't taste like much at all to her. Thus, students like Paula also learn that watching one's weight means giving up on certain food pleasures, especially those that they have come to enjoy from the tastier foods that their families cook at home.

In contrast to a focus on dieting and body size, my discussions with SGCP teachers and students revealed a quite different dynamic for male-identified students in regard to their food concerns and eating practices. As a whole, male-identified students tended to downplay lunchtime as an important event of the day, and were less inclined to talk about food in general. Interestingly, several boys from nonwhite, working-class backgrounds were particularly insistent that their focus on food while in school was driven by purely monetary concerns. Two boys, both friends from Central School, even told me that they often sold their lunch to make money, explaining to me that lunchtime was "business time" for them (Student, CA). The boys were able to do this since they qualified for a free lunch under the National School Lunch Program (see www.fns.usda.gov), and could sell their lunches at a discount to their peers. They also told me that they sold halloween candy, soda, and other food and nonfood items during lunchtime, all as part of this business endeavor. Of course, this profit-oriented positioning serves to bolster the social norm that men and food only mix when men are engaged in a professional activity. In Nova Scotia, a Plainville School teacher recounted a similar sort of dynamic, where boys distance themselves from the *act* of eating, though they will still engage in cooking for practical reasons:

Some boys think it's uncool to bring a lunch, like they are embarrassed to eat in front of people. For girls it has been a weight thing and for boys I dunno, almost like it's a girl thing to care about food. But they both cook it, because they get out of class to do it.

(Teacher, NS)

These social norms and behaviors are not just descriptive; they are also material. These boys were visibly animated when discussing the prospect of money-making through their acquired cooking skills, and their activity levels often waned when conversations turned towards dieting and healthy foods. Girls, on the other hand, often took to heart their role as care takers and healthy food advocates, sometimes chiding their parents for making grocery selections that did not conform to their school's version of "good" food. These experiences heightened their sense of empowerment and know-how, and motivated them to speak up when they otherwise might remain silent. While gender is clearly not the only lens through which we should seek to understand such students' bodily experiences of SGCPs, we would be remiss to not recognize how students' bodies are differently animated across lines of gender-normalized behavior patterns, as well as those of race or class.

As a whole, the above discussion suggests that despite many of my interviewees' insistence that gender was no longer really an issue for the SGCPs, students' experiences of these SGCPs were undoubtedly influenced by the gendered norms and expectations that continue to surround both food preparation and eating behaviors. The simple fact that both boys and girls are cooking the SGCP food does not mean that their experiences of the programs are the same. What viscerally motivates boys to engage in the programs was more often than not a desire to acquire what they see as professional skills, while girls were often inspired and animated by learning how to cook for their families and friends, as well as by the desire to engage in "healthy" dieting practices. These bodily impulses surely help to reproduce the traditional gender-based division of labor in the public/private sphere, as well as the gendered social norms surrounding body size and diet. But, as the next section will illustrate, these traditional impulses are also not the whole story.

Gender disruptions

I think me being a woman reinforces gender roles . . . but there are teachers in here who are men who set off light bulbs. That makes a difference.

(Teacher, CA)

Despite the obvious ways in which traditional gender roles are reinforced through SGCPs, it is important to recognize that there are also moments of disruption. These moments, although perhaps small, are not insignificant. For one, there *are* some male-identified teachers who do serve as role models for SGCP students, and who in my experience actively and consciously try to encourage students to think differently about who belongs in the kitchen. In Central School, for example, one of the students' male-identified homeroom teachers would frequently help out in the kitchen classroom, and would talk often to a group of male-identified students about how important the SGCP was. On one particularly memorable day, he engaged a group of boys in the task of crimping piecrusts for three pies. He explained that, "[his] grandmother had taught [him] how to do this, and [that he had] passed it on to [his] children" (Field notes, November 2007, CA). From the level of focus and animation within the group, it was clear that all of the boys were interested in learning how to do it, and they all watched this teacher intently.

At another time, this same teacher made a comment about the importance of "women's work" to a group of students (both female-identified and male-identified) that he was overseeing. Although the comment itself conformed to traditional stereotypes, the effect was notably empowering for the female-identified student, motivating her to be confident and take pride in the work that she was doing. The comment also called into question the devaluation of "women's work" in general:

Field notes, November 2007, CA

Mr. English made a comment about needing a girl to coordinate the table, partly because Sally [a female-identified student] was already taking charge and telling her peers what to do. He was trying to defuse the situation by showing that her strength and initiative was a positive thing. One of the boys was complaining that she was bossy, and Mr. English said in reply, "Try running your house for one day without a woman. It would be messy and disorganized." Sally responded to this with pride, and added that girls know how to cook and clean and do laundry too, implying that this was her space, and she knew how to manage it. The implication was that she had the right to take control within the group.

In addition to such attempts to (partially) disrupt gender norms among student interactions and practices, a few teachers also told me that the SGCPs additionally give the teachers themselves an opportunity to reflect upon and alter their own expectations and judgments in regard to gender. As one cooking teacher explained to me:

> The more I am able to open to these students, the more I reach them [the better] . . . so developing personal relationships is important . . . and over the years I have had a lot of judgments about behavior, especially with what I expect from the boys, like more misbehaving and disruption, and I have to get past it and be open . . . More and more I am developing those relationships with students. I am not saying I reach every child in that way, but we have moments, and for me that's a personal development, to be more open.
>
> (Teacher, CA)

Here this teacher suggests that through the SGCP, she came to understand that gendered behaviors are much more fluid and negotiable than she initially expected. In becoming more open to new possible outcomes, this teacher allows students some agency in the production of their gendered identities, and furthermore recognizes that her own stereotypical expectations were potentially limiting her students' agency in this regard. And indeed, students do have agency in the reproduction and resistance of gender norms. In my interactions with students, I found that standard gendered identities were frequently called into question in momentary but not insignificant ways. These disruptions often occurred, as Chapter 5 discussed, in the context of "play." For example, in a group interview with four boys, two students bantered back and forth about the importance of the SGCP to their lives:

> *Student K*: It's not a girl's job, it's just a job, everyone's job. I think it is important to have this class because when I grow up, I will know how to cook. If both husband and wife know how to cook, it'll get done fast.
>
> (Student, CA)

> *Student L*: Well, I think it is important because some of us aren't gonna get married [implying that Student K could never get someone to marry him].
>
> (Student, CA)

> *Student K*: Yeah, right, *some* of us aren't.
>
> (Student, CA)

Through a harsh but friendly back and forth between two friends, these students call into question the traditional role of the woman as domestic cook. While Student L implies that it is only important to know how to cook if one does not have a wife, the tone of this conversation was sarcastic. The boys remain comfortably within a heterosexual, masculine identity, but at the same time, the topic of their conversation allows for the possibility of something beyond traditional gender roles to emerge. In another small but not immaterial moment, three students were joking around with each other while chopping onions:

Field notes, December 2007, CA

Today there were a lot of onions to chop for the recipe (a bean stew) that we were making. One of the boys at table one made a comment that "Real men eat raw onions." He held out an onion for another boy to try, urging him to take a bite. A girl who was listening to the conversation blurted out, "No, men with real *bad breath* eat raw onions." Two other girls laughed at her comment. The other boy didn't eat the onion.

Here again students engage in disruptions of traditional gender roles through unstructured "play" in the SGCP classroom. This example is perhaps a counter to the bitter radish story, in which the SGCP garden teacher encouraged male students to be strong (in a fear factor type practice) and eat the spicy, bitter radishes. Here, a girl points out the absurdity of such a claim to masculinity, and the boy backs down from the act—either because he didn't want bad breath, didn't want to taste the raw onion, or both. It is important to note that, again, these are not just descriptive examples but also viscerally relevant as well. In the onion incident, the boy is challenged bodily to engage in something that supposedly male bodies should be able to handle, but that he clearly does not want to do. As the female-identified student disrupts this challenge with humor, the feel of the interaction shifts from aggressively playful to absurdly silly, and perhaps even a bit shameful on the part of the initial aggressor. The outcome, though small, is a collective sense that traditional masculine roles are to be taken with a grain of salt, so to speak— that the aggressive posturing is something to shrug off.

These small moments of disruption also happened in the garden as well as the kitchen classrooms. Although SGCP teachers sometimes claimed that it was the girls who disliked getting dirty in the garden, a field journal excerpt from an early day of planting in the Plainville garden revealed a different dynamic:

Field notes, May 2007, NS

The boys didn't want to touch the soil because they were grossed out by the horse manure. When I asked them to cover up the seeds, they kicked the soil with their feet rather than touching it. The girls from the same group knelt down and pushed the soil over the seeds. They told me that the boys wouldn't touch the soil because they were wimps. One girl then actually picked up some of the manure, which was in clumps, and dropped it behind her, saying "It's like . . ." and making a gesture like she was imitating a horse pooping. All of the girls roared with laughter, and their teacher shot disapproving glances at both me and girls.

Through unstructured "play" in the Plainville garden, these girls disrupt the gendered expectations of their teachers both in terms of who was more likely to get dirty in the garden, and also who was more likely to be a disturbance to the class. While the girls reinforce gender roles by calling the boys "wimps" for not wanting to touch the horse manure (insinuating that if they were more typically masculine, they would touch it), they also revel in the sense of strength that comes from engagement in an act that they experience as gender transgressive. If the boys are "wimps," then it is the girls who are unexpectedly strong and dominant in this moment—a fact that leads them to be noticeably animated (and perhaps also "disruptive" in the eyes of their teachers). Although their actions might have shocked (and perhaps annoyed) the SGCP teachers, leading to disciplinary action (an annoyed glance), the opportunity for gendered disruption itself is important to how the students come to experience the SGCPs. Unlike the various opportunities to engage in expected gender roles within the kitchen and garden classrooms, this incident illustrates that SGCPs can also provide opportunities for transgression.

Because this incident developed through a school garden activity, the gender transgression itself must also be understood as an outcome of the SGCPs, alongside and intertwined with any concurrent reinforcements. To be sure, the manure incident was one of many small moments that provided the potential for such transgressive outcomes within the SGCPs. In another complicated and contradictory example, several parents also reported being shocked, and often pleasantly surprised, when their *sons* particularly showed an interest in cooking. As one teacher explained:

> The feedback I get from parents is you know, my son or daughter, but often son, it shocks them because now their kid wants to cook. Parents respond by saying I have a child who loves this. Often the boys, they discover a new fun thing to do.
>
> (Teacher, NS)

This sense of shock is revealing because it illustrates what some parents were so adamant to forget—that men and boys are *still* not really expected to cook domestically, or at least to enjoy domestic cooking. Boys who cook were therefore more likely to receive praise from their parents and teachers, while girls who cook would often not generate as much of, or as positive of, a reaction. These experiences arguably reinforce traditional gender roles by conceding to male-identified domestic cooking as a pleasant surprise. Nevertheless, what this dynamic also tells us is that through engagement in the SGCP, some boys *are* learning to enjoy domestic cooking in ways that were previously unavailable to them, and in ways that could have repercussions for how they imagine and enact their role in household labor. Over the course of my research, the opportunity for boys and girls to both participate in cooking and gardening classes *together* was something that many parents agreed was not only a good thing, but was also something from which new understandings and new practices of gender identity could emerge. As one mother explained of the Berkeley SGCP:

> It's way better [than what we had to go through]. Home economics was very sexist, it was for house makers and you didn't see men in the classes often. Even in high school in the early 90s we had men in the class and they were there because they wanted the food not because they wanted to learn how to set a table.
>
> (Mother, CA)

In this example, the mother claims there has been a noticeable shift since she was in high school in the early 1990s, a shift attributable perhaps to the current ubiquity of concerns about obesity and healthy eating, as well as the alternative framing of the SGCP initiatives themselves. Undoubtedly, these programs *are* different than their 1990s counterparts—but how and why? The above sentiment suggests that one reason is gender. Gender roles are changing, however slowly, outside the walls of kitchen and garden classrooms, as well as within. As the SGCPs offer alternative spaces for students to enact (rigidly) gendered identities, so too do they allow opportunities for gender to be experienced as fluid and complex, even contradictory. As hands-on spaces that offer students a place to act and react outside the boundaries of most traditional classrooms, SGCPs can also provide a more flexible space in which traditional gendered identities surrounding food and cooking can be called into question and resisted. As the discussions above suggest, this resistance takes place not only through the actions of the students themselves, such as through their unstructured "play," but also through the actions and reactions of both teachers and parents.

In addition, such resistance can also take place as students carry their differently gendered food identities home with them. In contrast to the dieting-focused gendered behaviors that I discussed in the previous section, some students also articulated mixed reactions to the goal of dieting, expressing a more ambivalent relationship to the lessons of SGCPs, and to the overall goal of bodily thinness. Coming from female-identified students particularly, these reactions suggest that the SGCP students are not simply learning to accept and reiterate the valuableness of low-fat foods or thin bodies, but they also may be learning to think critically about these values and to reassess their import. In asserting disagreement, the female-identified student below registers not just her ambivalence about the SGCP lessons overall, but also her resistance to the idea that women should primarily be concerned with healthy eating, and should adhere to the logic of SGCPs.

> I mean, I still eat healthy and everything [at home], except I do use butter and cheese instead of oil and soymilk. It's not like I eat fast food or anything. I mean, [the SGCP] does have an effect on what I eat at home, but I don't try to do everything the same.
>
> (Student, CA)

While this student draws a hard line at fast food, she also recognizes that her food preferences (e.g. butter and cheese) matter to her enough that she isn't going to follow everything that the SGCP tells her is "better," even if she believes it to be healthier. Rather than simply doing "everything the same," she is able to take the lessons that the SGCP offers her and weigh them against her own ideas of what is good to eat. Other interviews with students and parents, as I have noted, recount much less ambivalence towards SGCP lessons, where female-identified students particularly take the lessons very much to heart. While these examples do illustrate a gendered focus on healthy eating, however, they also highlight potential shifts in other gendered relationships. For example, a father recounted to me the changes that he saw take place in his relationship to his daughter when she enrolled in the SGCP classes at her school. He was proud that his daughter had taken up the cause of healthy eating so vigorously and knowledgably since she began the program, and was happy that the lessons had made her speak up—even when it was against him! Referring to foods that he and his wife choose for their household, the father explained of his daughter:

> She gives us feedback at home [now], and if [our food] it is not very healthy, we hear it! [She'll say,] "That's too fatty," or something like that; she is very conscious of what healthy is and then that is fed back to us at home.
>
> (Father, CA)

Although this student clearly adheres to gendered ideas of healthy eating (particularly in terms of the policing of fatty foods), she also does so in such a way that, according to her father's account, has shifted the power dynamic within her family and empowered her to make and defend her food decisions in direct contrast to her parents' food habits. While this particular example is seemingly small and mundane in relation to much larger power imbalances that exist beyond the household, we cannot know what this small shift in power might mean for the student's ability to speak out about other life choices, to question the decision-making qualifications of experts or authority beyond her parents, or to be generally confident in her own decision-making abilities. Indeed, it is through food choices that young people can often start to discover their agency in political decision-making— for example, by becoming vegetarian or vegan, or by eating in ways that counter what they grew up with. In this sense, SGCPs can also open opportunities for such students to learn significant lessons about agency and empowerment, especially in relation to the influences of hegemonic gender norms.

Concluding thoughts

In all my interviews, focus groups, and discussions with research participants, on no topic was there as much agreement; almost everyone I spoke with insisted that gender was no longer relevant to issues of food. "Both boys and girls can cook now," the parents and teachers explained; the students insisted the same. And still, a large percentage of those who I met over the course of my research were women—mothers, teachers, food activists and cooks. For this reason alone, I repeat the question, how could gender *not* still matter? This chapter has explored the continued importance of gender identities to the varied experiences of people engaged in alternative food activism. It focused on stories that illustrate the hidden realities of gender, and the ways in which gender identities can influence seemingly individual food desires and practices, even as people insist otherwise. Some of the stories in this chapter have also demonstrated that gender norms can be overt at times; and, importantly, it is during such overtness that gendered reactions and practices can render themselves especially challengeable and changeable.

Given the salience of gender to the students' (and adults') experiences of the SGCPs, the teachers, parents, volunteers, and leaders who are associated with SGCPs could certainly approach the question of gender difference more openly and honestly. Rather than insisting that it is largely "a thing of the past," these adults could help the SGCP students to be cognizant of the gender identities and expectations that they bring with them into the kitchen and garden classrooms. They could help them to understand how motivation to cook or eat is dependent not just on what they might consider to be "individual preferences" or "tastes," but also on how students are socially and culturally compelled to conform to gendered norms and patterns. Moreover, teachers and leaders could also think of ways to promote creative transgression of these gender boundaries, rather than leaving the disruption largely to unstructured play and happenstance. Nevertheless, despite these coulds, the cooking and gardening programs that I examined do seem to provide children with a relatively open space for gender disruption that might otherwise not exist for them. These SGCPs give their students a unique, school-based opportunity to collectively explore, (re)act to, and (re)create their gendered food identities.

7 We run it all off!

They want us to learn how to eat good. You know how a bunch of kids will usually eat hot chips and stuff? And so they want us to make good decisions . . . The only thing is, yeah we eat food, like junk food, but we run it off, we run it all off.

(Student, CA)

"What do you think the purpose of your cooking and gardening classes are?" I asked the above-quoted 7th grade student as we headed towards the garden. His response was at once simple and loaded. "Yeah, we eat junk food," he explained, in knowing contrast to the good food that "they" want him to eat, "but the thing is, we run it off. . . we run it all off!" This sentiment encapsulates a typical moment in the daily, school-bound life of a middle-school aged student in Berkeley, California, as he navigates the interconnected worlds of dieting, health, exercise and foodie-ism. None of these adult-driven worlds—the worlds of "they" the experts—fit neatly on to the terrain of his youth, despite the fact that almost all middle-school children can quickly rattle off a slew of one-liners to please their teachers.

When asked by an adult what they think of their school garden or cooking classes, students would frequently articulate answers that they assumed the adults want to hear: "The garden makes us healthy," "Fresh fruits and vegetables taste better," "It's more fun to eat what we grow." These are phrases that the students have been taught are "correct," and are the outcomes that most adults in the programs are hoping and looking for. They are not necessarily, however, what middle-school students actually believe to be true, nor what motivates them to engage in particular food behaviors. This chapter uses youth-driven narratives to explore the realm of kid food, a realm that emerges not from the sleekly mapped ideas of what adults assume kid food to be, nor from the neat one-liners through which kids articulate what adults want them to say, but from the varied experiences and ambiguous accounts of the middle-school aged children themselves. What becomes clear, once again, is that bodily motivations to eat certain foods can function to both reinforce and resist the boundaries of social difference, including those of age and authority.

Agency in food-based decision-making is complicated when talking about children. The common societal belief that food choices are ultimately made as rational, individual choices often means that responsibility to make the "right" decisions falls upon the person who buys or consumes a particular food. This is not usually the case, however, when discussing children's food choices. For example, in marketing and public health documents, children are often deemed to be an *irresponsible* party, while the accountability and agency falls instead upon their parents (Colls and Evans 2008). Historically, a key target population for encouraging nutritional change has been low-income young women (presumably as current or prospective mothers), with the goals of change focused on their individual consumer choices rather than any structural inequalities that these young women may face (Slocum et al. 2011). In the popular reality TV series *Jamie Oliver's Food Revolution*, the show addresses viewers as individual parents with a moral obligation to feed their children correctly—parents who should feel shame when their children make bad food decisions. Although the TV show positions children as important food decision-makers, it also recognizes that letting kids make their own food decisions is "probably not a good thing," and admits that adults must work to convince children of what is actually "good to eat" (Zimmerman 2013).

In regard to SGCPs, program leaders and teachers seem to frame children's decision-making abilities in a variety of ways. On the one hand, adult program participants did often assume that children have some amount of individual agency in making or controlling their food decisions—both inside and beyond the boundaries of the school itself. Indeed, teachers and leaders of the SGCPs frequently repeated phrases like "children are the seeds of change," (Teacher, CA) indicating a hope and faith in children's ability to shift the "bad" or "unhealthy" food choices of their families and communities. In this sense, the programs looked to children as important leaders and change-makers, although the direction of their leadership and the changes to be made are still defined by adults. On the other hand, as we saw in earlier chapters, parents are also often blamed for bad parenting abilities when their children make poor food decisions, or when their children look visibly "unhealthy" (read: fat). In addition, SGCP teachers and leaders sometimes assumed that children are not "in control" or "responsible" enough to eat healthy foods without some amount of trickery on the part of adults—for example, blending vegetables to hide them from other students in various soup and pasta sauce recipes. Furthermore, SGCP teachers and leaders often insisted on the importance of some amount of mood control in their classes, such as the rule that students can only say positive or neutral statements about the foods they are eating, and must avoid making negative reactions that could influence their peers. Indeed, Alice Waters' vision in her "Delicious Revolution" of "seducing" children such that they become "enraptured" by vegetables assumes that children's food choices are not informed by rational decision-making but instead through an instinctive and animalistic desire for sensory pleasure that must be harnessed by adults in order to captivate and craft them as eaters of "marvelous" foods (Waters Online). In all of these examples, children again become irresponsible and incapable, while individual adults (most often, mothers and teachers) become those who are (or should be) both responsible and culpable.

For all the above reasons, it is important to examine the various ways that SGCPs might conceive of and (re)produce students' agency as consumers of food. How are children positioned vis-à-vis adults and the wider social community in terms of their respective responsibilities to make food decisions? What assumptions are made about how children, in contrast to adults, are able to make (good) food decisions? And how do these positions and assumptions inform the way that the SGCPs seek to motivate children to make healthier choices? In asking these questions, it is particularly crucial to evaluate the notion that individual choice ultimately drives children's food decision-making, since as we have seen in past chapters, varying socio-economic and material conditions certainly shape the contexts in which any one individual can act "responsibly." In addition, however, it is also important to acknowledge the ways in which children are able to negotiate and (re)practice their own personal food-based identities in respect to their relative youth. The sections below discuss how the SGCPs in question offered students important opportunities to practice their agency as food decision-makers, at the same time as the programs reproduced some unhelpful assumptions about the ability of children to make "responsible" food decisions.

Kid food

> We couldn't get anyone to listen to us, so we said, let's start with the kids, they are a captive audience.
>
> (Leader, CA)

> Working with kids, we have a captive audience, they don't have a choice, and you are making a change early, so hopefully the benefit will last a lifetime.
>
> (Teacher, NS)

Consistent with the enlightenment frame of SGCPs, the SGCP teachers and leaders typically positioned children as inactive recipients of food education, who are easy to reach because they are a "captive" audience. In the above quotations, students are envisioned as passive tools of the alternative food movement; they are empty vessels that can be filled with information (regardless of their consent), and then used as vehicles for passing along this information to parents. These quotations are typical of the way that many SGCP teachers and leaders described to me the role of children in promoting alternative/healthy food, locating kids as passive participants in the production of food knowledge. As many of my interviewees noted, the role of children in these SGCPs is similar in this sense to that of the anti-smoking campaigns of the 1980s, in which health education in schools was meant as a way to reach adult smokers. In this understanding, change takes place only by reaching the truly *responsible* party—that is, the parents:

We started talking about the language of healthy eating and we could put the words in the kids' mouths. So, when we asked, why are you eating that? They respond, because it is healthy and prevents diabetes, and it is good for my heart, or whatever. We put the words in their mouths. And then we got the parents, because if you give the kids a language, it becomes a reality. The kids took the language home, and it became a real cultural shift.

(Leader, NS)

In this explanation, the SGCP leader assumes that children have internalized food preferences simply because they know the "right" words to say about what and why they are eating. This SGCP leader imagines change to occur through a process of filling students with the "correct" language and then encouraging them to repeat this language at home. While it is certainly true that language and rhetoric can lead to cultural shifts, this reliance on a linguistic framing also reinforces the idea that knowledge-based solutions focused on the education of individuals are the key to food-based social change. Moreover, it also assumes that because students know what adults want them to say, this must also mean that students will do what adults want them to do, and feel what adults want them to feel. In other words, it assumes that a language shift means that the programs are "working." In my interviews and daily interactions with students, I came to realize that many students were indeed conversant in the language of SGCPs. Students were acutely aware of what it was that adults wanted to hear, in addition to what adults wanted them to look like (as the opening quotation suggests). But, whether this awareness meant that students actually believed what they were saying is a different matter altogether.

Students' ability to say what was expected of them came through very strongly in my interviews with many students, when I asked them why the SGCPs are important to them. I very quickly came to realize that my student interviewees already knew the correct sound bites to repeat, and that, since so many adults had praised their ability to repeat these sound bites in the past, they assumed that this was what I was also out to hear. "This food is local" (Student, NS). "It's fresh and healthy" (Student, CA). "[The program] teaches us how to make better choices" (Student, CA). These were all phrases I heard repeated again and again by a variety of different student interviewees. But, as I quickly realized, the ability to repeat such sound bites does not imply that these students actually believed what they are saying, or that they had actually taken it to heart. The parroting only showed me that they are quite good at repeating the information that is "poured" into them, as a "captive audience," but it does not confirm that the program is "working." As one student told me to the contrary: "It's not really working; kids still eat junk. I mean, it's all good, but the kids will still go to 7–11 and buy their hot chips. Last year there was a big group that would go every day to 7–11. I only go once and a while" (Student, CA).

If simply convincing kids to eat healthy, alternative food is the ultimate goal of these SGCPs, as this student implies, "it is not really working," at least for some students. One reason for this failure to translate words into action might be that the SGCP leaders and teachers tended to give little credence to the ability of students to think about and (re)act to food on their own terms. For example, students go to 7–11 to buy hot chips not only (or even primarily) because of the types of foods there (i.e. healthy or not), but for a variety of reasons related to their social position as youth: it is a walkable location, it is accessible by students without the presence of an adult, it has cheap snack foods that many kids can afford, and it is a designated hang out space for many middle-school aged kids. The decision to go to 7–11, though not "right" in SGCP terms, is therefore in many ways a thoughtful and logical choice that illustrates that children are capable and active decision-makers. In this sense, the insistence that the programs are working simply because of linguistic changes implies a lack of consciousness about the ways that students navigate and make decisions within the various food worlds that surround them.

As empty vessels filled with linguistic information, students presumably become instruments of food-based social change—entrusted with the task of carrying forward correct food knowledge and behavior. In this way, their capacity for good decision-making is acknowledged even as their agency is limited by adults' ideas of what constitutes a good decision. At other times, however, SGCP leaders, teachers, and even parents were insistent that children were simply not capable of making good decisions, and that adults were the only truly responsible party. One such way is through the trend of tricking kids through hidden vegetables, as seen in cookbooks like *Deceptively Delicious* (Seinfeld 2007) and *The Sneaky Chef* (Lapine 2007), and as repeated frequently in the kitchen of my Nova Scotia case study site. The assumption behind this trend, as many teachers and parents expressed to me during interviews, is that children will not eat healthy food if you tell them that it is healthy. The only way to get them to eat food that is good for them, then, is by tricking them into thinking that it is not as good for them as it actually is. Vegetables—the quintessential good food—therefore need to be hidden (through blending, or fine chopping) in order to convince children to eat them. Of course, as some SGCP teachers acknowledged, this practice backfires when you actually teach students how to do the hiding:

> I think sometimes some of the kids love the food, like spaghetti or pizza or something, and then they go in there [to the kitchen] and they realize that there is a green pepper in it or something, and it is like, I am not eating that anymore. It is funny because you were eating it one day, and nothing has changed except now you know what might be hidden in the sauce, so it is all a mind thing sometimes.
>
> (Teacher, NS)

This teacher's explanation is a complicated one. She acknowledges that many kids like the taste of certain "healthy" foods, but also that they do so because they don't know the specifics of how these foods are made. Once the students know the "truth," she laments, they sometimes refuse to eat the same foods that they used to enjoy. We could analyze this story in a number of ways. On the one hand, as many parents would attest, it is hard to please picky or timid eaters. Children frequently assume that they won't like certain foods based on certain ingredients, but then end up actually enjoying them. Given this, the teacher assumes that her challenge as an adult is to trick the students into eating foods that they might knowingly refuse, engaging them in a mind game of sorts. In this sense, children become an irresponsible party, not capable of making reasonable food decisions. On the other hand, we might also acknowledge that the children in this story could likely be reacting to the experience of deception itself. That is, their refusal to eat the food that they previously liked comes not from irrational and irresponsible decision-making but as a logical response to the realization that adults had been trying to dupe them. This assertion of students' agency over the adults' attempts to manipulate their eating habits seems less a part of a mind game than a viscerally driven effort to stake claim over what enters their individual bodies. My conversations with several students about the "hiding vegetables" issue seemed to lean towards this second explanation, as they expressed shock and dismay over what felt to them as a betrayal of trust.

In addition, however, the above story also begs the question of how we determine (or predetermine, as the case may be) which foods are likely to be kid friendly. The rhetoric surrounding the notion of "kid food" stems from a belief that there exists a binary of food types, kid foods and adult foods, and both are mutually exclusive and naturally determined by age. Healthy foods, including especially vegetables, seem to fall in the adult foods category as something that only rational, responsible people (read: adults) would choose to eat. Junk foods, including candy and pizza, more typically fall under what is considered by many to be kid food—foods that are instead chosen through an irrational or irresponsible desire that is driven solely by taste buds. Both children and adults repeated the rhetoric of kid food, and both children and adults also performed these age-specific expectations, despite the fact that what kids and adults actually like to eat is surely more fluid and overlapping. The question of what, then, determines the anticipated boundaries of kid food, and what solidifies these boundaries into food habits and desires, is a question that is only answerable in particular spatial and temporal contexts. Within the context of the particular SGCPs in question, it seems, kid food comes to matter both as temptations that students need to be saved from, as well as disguises that can trick students into liking foods that are healthy and thus undesirable.

What was most striking to me about the rhetoric of kid food in SGCPs was the way in which the rational mind appeared to be written out of the decision-making process. SGCP teachers and leaders often imagined the hands-on, material lessons of the SGCPs as mechanisms through which to tap into the students' precognitive bodies and train them, or trick them, into doing what is right. This idea of a lack of conscious, rational decision-making on the part of children stands in contrast, one would think, to the idea that students can become literate in the language of healthy eating, learning to repeat the rationale behind why eating healthy foods is important. However, this repetition was more often perceived as a way to reach adult parents, and achieve nutritional change at home. True, bodily change on the part of children alternatively seemed to require a more embodied methodology, according to some interviewees. As one mother explained, when considering the utility of her children's SGCP in convincing them to enjoy eating healthily: "I think Alice Waters is a bit out of touch, but she is also right; you can't have the kids eating healthy by telling them broccoli is good for them. You have to have them fall in love with broccoli" (Mother, CA).

While it is important to approach foods in a variety of ways, and not only through intellectual discussion (or a lecturing monologue, as the case may be), this assertion raises the question of why we think that we cannot actually talk to children about food. One answer is that adults often assume that kids operate on a more bodily level, driven by visceral drives for pleasure or fun, which are considered in opposition to the cognitive work of the adult mind (Colls and Evans 2008). Hence, we see advertisements for kid food and kid friendly spaces, which are usually brightly colored, creatively named and covered with "fun" games and activities. But are students really disinterested in talking about food, or rather are we (adults) disinterested in hearing what they actually have to say? I would venture to say the latter is more the case than the former, especially when what children have to say about food is not parroted sound bites but perhaps more ambivalent or even negative reactions to healthy eating.

The practice of not talking directly to kids about food (at least in ways that might be perceived as negative) also translated into other absences within the SGCPs that I studied. For example, the politics of alternative and conventional food was rarely if ever discussed in the SGCP classrooms, particularly because politics were seen as too controversial and potentially off-putting or demotivating to students, again because of the possibility of negative visceral (re)actions. Instead, leaders and teachers advocated a "positive politics" or "politics by osmosis" approach, whereby simply showing the kids how fun food can be (and avoiding all negativity) would lead to changes in the students' food behaviors and preferences. In addition, the social pressures that surround weight and body image were also rarely discussed in the SGCPs that I studied, at least not directly and purposefully by the SGCP teachers and leaders. As one dietitian instructed me: "You *never* mention weight with kids, you *never* mention their looks. You don't want them to feel guilty" (Leader, NS).

This sentiment stands in stark contrast to the opening quotation, in which the student assures me that eating junk food is not a problem because the kids just "run it all off." Through many conversations with students, it became clear to me that many students considered the SGCPs to have something to do with "their looks," regardless of how often this topic was discussed by their teachers. Indeed, students came to their SGCP classrooms with clear ideas of the connections between healthy eating, body size, and beauty, as defined through popular culture, social media and even medical discourse (on the importance of avoiding obesity, for example). They also undoubtedly came with many different, culturally specific ideas about what sized bodies are actually desirable, and how much weight is considered beautiful. And, many surely came to the SGCPs with a variety of plots and proposals for how they might use the SGCP lessons to achieve their own body weight and appearance goals. To not then take up these dilemmas and discuss them directly with students, it seems, is therefore to deny students' experiences and expertise in navigating these complex questions. It is to consider students as simply "empty vessels" to be filled with (only) positive information, as defined by SGCP leaders and nutrition experts, but never by the students themselves.

Taken as a whole, this approach to food education seems questionable, and perhaps counterproductive, to encouraging students to develop a healthy and positive relationship to food. Some SGCP leaders and teachers engaged in activities that "trick" kids, or attempt to "seduce" them, and that ultimately seek to use them as vehicles to get to a more important and responsible party—their parents. They avoided talking to children directly about the negative aspects of food because they assumed that children cannot handle such discussions, or that they will be turned off to healthy food as a result of hard conversations. Yet, at the same time, the SGCPs had a very particular set of "rights" and "wrongs" when it comes to food. They encouraged children to learn these rights and wrongs, and indeed to internalize them, but they did not encourage children to take an active role in the production of such food knowledge, and they did not seek to understand the diverse food worlds that children actually inhabit (at least from what I saw). In these ways, the SGCPs reproduced unhelpful assumptions about the (in)ability of children to make "healthy" food decisions, and also about the ways that children's (bad) food decisions are driven (more so than adults) by irrational bodily desires.

Adult tastes

Despite the fact that, for all of the reasons detailed above, some SGCP leaders and teachers did not actively encourage students to find their agency within the processes of food decision-making, many students found and asserted their agency regardless. The lived space of the SGCP classrooms was one of many social spaces in which the students could find an opportunity to do so (in addition to spaces like 7–11, the cafeteria or their homes). In some cases, the students' agency emerged as a disruption of a planned SGCP activity; at other times, students' agency came through their engagement with the ideas and information that the SGCP classrooms provided. During my participant observation in the SGCP classrooms, I frequently witnessed students engaging in disruptive or dissenting activities that served to underscore their position as active participants in the processes of making and eating food:

Field notes, November 2007, CA

None of the kids really liked the salad that we made at all today. Ms. Tiffany and I both said we liked it. But Melissa asked the other students if they liked it, and no one said yes. She told us that it was an adult taste, and that they should have been able to alter the recipe to their liking (specifically, to leave the dressing off the salad). Ms. Tiffany had previously said that the students should use the dressing as it was specified in the recipe, and that they should try it like that. But, later she said she felt bad that she didn't let the kids leave it off, and during the next class she let them keep the dressing on the side.

In this example, a student named Melissa uses her social position as a youth to question the decision-making ability of the adult authority figure, Ms. Tiffany. Melissa draws on the notion of adult tastes to suggest that what tastes good to an adult authority figure might not taste good to the rest of the participants. Although Ms. Tiffany does not immediately accept this proposition, she eventually responds by changing her own behavior in regard to how she approaches the instruction of her next class, suggesting that Melissa's critique of her food choices was both legitimate and effective. Notably, Melissa draws upon the kid food/adult taste dichotomy as a micro-political strategy that allows her to assert her taste preference as legitimate. Other students beyond Melissa also drew upon this dichotomy in asserting taste preferences, although sometimes in different ways. For example, when I asked one student about whether she liked the foods they cooked in her cooking class, she replied: "Sometimes, [but] some things we make are nasty (no offense). I would rather have something not cooked by kids but [rather] by adults" (Student, CA).

Besides this kid food/adult taste dichotomy, I also witnessed many students using another youth-based food identity—the "picky eater"—to similarly assert their authority in the SGCP cooking classroom. The picky eaters of SGCP classes were often well known by all members of the class, and widely acknowledged as hard to convince; they could use their identity either to remain hidden and distant from the eating activities of the class, or to garner the attention of teachers and classmates. Many SGCP teachers were certainly conscious of how students used such identities to gain authority and agency in the classroom. While some teachers found this dynamic unhelpful and disruptive, other teachers understood that students used these identities to carve out a space where they could feel like "themselves" within the SGCP classrooms, and indeed within the broader food system at large. These teachers also saw the role of SGCPs as one of providing students access to other kinds of identities beyond picky eating:

> Yesterday we had a girl say, "I don't eat vegetables," and she had a plate of vegetables in her hand. So I said, "That's interesting, what do you have in your hand right now?" And she said, "Well, I mean I usually don't." I said, "Well, I hope you are changing your image of yourself because you certainly look like someone who eats vegetables to me!" But, you know, some of them have these sort of identities around food; I am the one who never eats the skins on the apple, or this and that. So [the SGCP] gives them other ways to think [about how] they could identify themselves instead of being the picky eater, because they are always looking to be noticed. The picky eater gets noticed.
>
> (Teacher, CA)

If we analyze the information and ideas that emerge from the SGCPs in terms of opportunity, then certainly the SGCPs do provide students with access to other identities and knowledges that can expand their visceral attachments to food well beyond picky eating. When students are included as active participants in the process of creating food knowledge, the "shoulds" of the alternative food movement become instead bits of situated information that can evolve into discussions and "teachable moments" within the SGCP classroom. In this sense, the SGCPs become spaces where students can learn to assert their agency:

> [It is] not so much of a judgment thing, [it is] just information [that SGCPs give them]; like you guys need to understand that you are being targeted as consumers . . . and I talk to them about these little things, like how do you feel when a store says no more than three students at a time? What does that sign mean, you know? Do you want to support people who think you are a thief? . . . Or [do you want to start] demanding better food choices [in these places?] . . . so there is a lot of politics to it but it is sort of just something that is more a teachable moment that might arise, not something [planned ahead of time].
>
> (Teacher, CA)

Many students expressed to me how important it was to learn the information that they did within their SGCP classrooms. Although some did not agree that SGCPs were "empowering" to them, as SGCP leaders claim, many students felt that SGCPs did increase their ability to effectively navigate their own food worlds outside the bounds of the school. Certainly the knowledge that students gain from their SGCPs influences their ability to make food decisions in the broader world—both in their home spaces, and in other social spaces beyond the home. In this sense, SGCPs encourage students to assert their agency within these broader spaces. Some of the parents that I talked to acknowledged this effect, suggesting that it is actually the children and not the adults (the parents) that really hold the knowledge about what and how to eat:

> Ask the children, they know and they will tell you, [if] you put the thing in the wrong bucket, [they will tell you]. And, I think that is how this garden thing is going to work, if the kid knows, and the kid is telling the mother, [you] know what, we could have a baked potato [rather than fries], that is another way of having a potato. [So] the kids . . . that is where the power is, [they are] little dictators.
>
> (Father, CA)

Although the above quotation could be read as another story of kids, as empty vessels, bringing information home to their parents, in this description the children are the ones who are framed as active and assertive; they are said to hold the power. To not recognize the (at least partial) agency of the kids in these moments is to deny that they have the capacity to consciously and creatively effect change. One of the most important ways in which this assertion of agency comes to matter in the broader world is through children's disruption of the notion that kid food is equivalent to fast or junk food. Indeed, it is important to recognize that beyond hidden vegetables, another way in which students might be "tricked" is through the advertising industry. For example, scholars have discussed the ability of the McDonalds label to "trick" kids (as well as adults) into buying fast food products (Robinson et al. 2007). As my interviews revealed, many SGCP teachers and parents also often worried about the marketing of fast food and processed foods towards kids, and are compelled to find ways to get kids "away from [such] influences" (Mother, CA). In this sense, it is significant that the SGCPs can function as a space in which kids are encouraged to question and shift the association between kid food and fast food: "Outside of school, there is a lot of unhealthy stuff. There is preservatives in food, like McDonalds—a lot of kids go there to eat. It shortens your life. I only go once in a while" (Student, CA).

Here, this SGCP student is able to make connections between age-specific behavior patterns and the fast food restaurant McDonalds. He articulates particular health concerns—related to preservatives and the shortening of life expectancy—and critically reflects on his own behavior as a "kid" that resists such patterns in light of this health information. These sorts of discussions were not untypical as I discussed with students how they used SGCP lessons to navigate their food choices outside of school. Frequently, students articulated a critical and even skeptical position regarding their relationship to age-specific food behavior patterns.

Beyond the damaging impact of the fast food industry on the notion of "kid food," some interviewees also discussed how the SGCPs could give particularly disenfranchised students important opportunities to move beyond the economic and social limitations of their families and communities, particularly those that are related to nutrition and bodily health. To be sure, this was envisioned as a key function of these programs, and was one of the main reasons that they can rely on government funding:

> [The program that helps to fund the SGCP] is a program that is funded through the USDA food stamp program, and the monies are dispersed through the CA department of health. They disperse the money to schools if they have over 50 percent eligible for free or reduced lunch. It is geared towards low income because they experience disproportionate impacts in regard to diabetes and obesity and also poor food security.
>
> (Leader, CA)

Both in California and in Nova Scotia, the SGCPs were frequently positioned as programs that can particularly encourage or empower disenfranchised children to counteract the disproportionate problems they face in regard to dietary disease and food security. But how they can, and do, work to achieve this is a broader and more complicated question. More specifically, in what ways might we see SGCPs as providing opportunities for students to locate their agency in responding to and navigating through particularly tough food worlds? As one teacher articulated, the opportunities that the SGCPs offer disenfranchised students are not only intellectual but also embodied as well:

Now, as I snip my green beans and put them into the steamer, so that I keep in all the nutrients, I can remember never having fresh vegetables, I can remember taking them out of the freezer and boiling them until they had no taste, and wondering why. And so for me, [eating fresh green beans] is an experience that I can value and appreciate. As a young person [like many of my students], I couldn't make those connections.

(Teacher, CA)

In this teacher's description, the SGCPs exist as an opportunity that allows students to make the connections that he as a child, by way of his family's social position, did not have the opportunity to make. By intellectually and viscerally encouraging students to draw a relationship between bodily health and food choice, this teacher argues, the SGCPs encourage shifts in individual eating that reflect students' agency to move beyond the limits of their social positions. Moreover, by making these connections, he explains, students can come to feel good about eating the fresh green beans, and can even come to enjoy them. In this way, students' bodies are enrolled in social change not through trickery or mind games (as in the examples from the section above) but instead through a conscious and deliberate process of intellectual and sensory stimulation. Similar to this sentiment, one low-income student from Berkeley explained to me that the garden program provided a new experience that she deemed especially important, despite the fact that many of her peers complained about the garden:

JHC: Would you be upset if they stopped having a garden class?
Student: Well, I think [the garden] is nice because all the places I've moved, I ain't never had a garden. We didn't have space to plant stuff in my yard.

(Student, CA)

In this quotation, the student offers an unusually favorable interpretation of the garden in particular (a space that many students disliked in relation to the kitchen). She implies that the SGCP helps her specifically by giving her entry into a physical space that she otherwise would not have access to, either for economic reasons, or reasons of time or family priority. In this sense, we might say that the SGCP helps to disrupt structurally uneven tendencies of access, which are obviously physical in nature (providing the student access to a new physical space), but also visceral as well (encouraging the development of a new, embodied sense of spatial enjoyment). By giving such disenfranchised students the opportunity to relate to and form visceral associations—both positive and negative—with school gardens and with the produce that is harvested from them, the SGCPs enroll students in an embodied project of food-based social change. Importantly, they do so not only in prescripted ways (such as kid food or adult food) but also in ways that can encourage students to find agency in the processes of nourishment, such as when the above student goes against the grain of peer behavior to reveal her enjoyment of the garden.

This allowance for multiple reactions and feelings to emerge is important, and is something that I take up again in Chapter 8 in reference to food-based pedagogies. But, it was also something that was important to my own interpretation of what was happening in SGCPs, and how students particularly were involved. When I began to look beyond the most obvious metrics for determining whether or not the programs were "working," I began to see that there are many ways in which to evaluate their work. For example, consider the following quotation:

> *JHC*: Did you enjoy yourself [in cooking class yesterday]?
> *Student*: I liked getting my hands in the bowl, there were spices, and you had to mix them all up together, the potatoes with the spices, and some oil, my hands got all oily.
>
> (Student, CA)

In this quotation, the student offers a description of a cooking activity that felt good to him. Notably, this description does not necessarily imply that he ate or enjoyed the food he was making, but rather that a number of different experiences and associations were formed through this particular SGCP experience that go far beyond the simple "outcome" of food choice. In this sense, and even as the programs are evaluated along more rigid and predetermined lines (e.g. changes in children's food behavior), the SGCPs events and spaces themselves allow students to relate to food in a variety of ways that push the boundaries of food–body relationships beyond the simple act of eating "good" food. Consider the two statements from students, the first offered as a description of a meal eaten at home, as part of a homework assignment, the second as an overall assessment of her SGCP from a letter writing activity undertaken by a student in Nova Scotia writing to another student in California.

> Zachary's Pizza prepared [my dinner] and I brought it home. I felt calm [eating it] because I was on the soft couch.
>
> (Student, CA)

> In the kitchen, we cook potatoes, grilled cheese, macaroni and cheese, and lots more. I don't like to peel stuff or spend my free lunchtime in the kitchen. Working in the garden is sorta fun. I hate the part where you have to get dirty, but at least we get away from class. We don't go to the kitchen as a class, just in small groups. This year, we built a composter from wood. It was REALLY fun. I like to drill.
>
> (Student, NS)

The first statement offers a brief description of a dinner meal in which the student ate pizza—a food generally understood by SGCP students to be unhealthy, and as such, a food to avoid. The student's description of his meal is notably unapologetic, and further, expresses a body-centered rationale for why this meal was particularly important to him, and why it felt "good" (calm, and soft) despite the fact that pizza does not fall into the "good" category. I include it because it strikes me as a particularly good example of real, honest body literacy—the ability for students to move beyond the sound bites that teachers might want to hear and articulate instead experiences that matter to them. In the second statement, the student articulates both positive and negative assessments of her SGCP. She doesn't like to peel stuff, and would never choose to spend her free time in the kitchen, but she seems excited by the variety of recipes she gets to cook. Her garden experiences are similarly ambivalent—the dirt is something to be avoided, but the sense of accomplishment in the act of drilling was something to remember fondly. These articulations go far beyond the sound bites once again, expressing a much deeper and more complicated relationship to the SGCP.

It is important to highlight that most of my interviews with students, as well as many parents, teachers, volunteers, leaders and activists, were filled with such moments of contradictory feelings and perceptions in regard to SGCP initiatives. This was true not only between different interviewees, but also in terms of the ideas and sensations expressed by each individual participant. In fact, the female-identified student above who spoke favorably about the school garden was also one who complained frequently about getting her shoes dirty and feeling too tired to work in the garden. (Later in the interview, I asked her, "Wait, I am confused, do you think the garden is important or don't you?" to which she replied, "Yeah, it is, we eat that shit [the produce].") Likewise, the boy who implied the program was irrelevant because the students "run off" the hot chips anyway was also an active participant in the cooking class, a class that he had elected (for two years) to take. In this sense, the data that I collected from interviews does not help to clarify why a particular student makes a particular choice, or how a specific social position configures a specific (re)action. Indeed, clarity is not what these interviews provided me. Nevertheless, such data did help me to see not only the complexity of food-based motivation, but also the different kinds of opportunities and moments of relating that SGCPs do and can allow. It is in understanding these opportunities and moments that we can (as teachers, leaders, activists, and academics) begin to rethink and reconfigure food education in ways that nudge forward an ever-more progressive visceral politics.

Concluding thoughts

Field notes, December 2007, CA

One kid today was telling me how he put some extra pepper in the recipe to spice it up a little more and make it taste good. He asked [the cooking teacher] if he could put a clove of garlic in and was pleasantly surprised that she said yes. He told me that he was going to put a little more olive oil too, and did so without asking.

In the quotation that opens this chapter, the 7th grader explains, "*They* want us to eat good," and "*they* want us to make good decisions." In his explanation, it is clear that the experts are a "they," and that this "they" does not include students like him. To the extent that the outcomes of SGCPs are dictated and measured by expert knowledge, we might say that SGCPs are limiting to the students' ability to find their agency within the SGCP classrooms, as well as within the broader food system. Yet, as my above field notes excerpt describes, this is clearly not the only way to analyze the outcomes of SGCPS and their impact on student empowerment. The micro-politics of student–adult relationships in these SGCPs are much more haphazard and contradictory than the expert-focused narrative allows. For example, in the above field note excerpt, the student clearly asks an authority figure—an adult teacher—for permission to alter the recipe that he was working from. However, before asking, this student had already taken it upon himself to try something a bit different by adding some extra spice! In receiving permission for additional alterations, and being affirmed by the authority figure, this student then goes on to alter the recipe in a more significant way, this time by adding more oil to his creation (something that was often read in the SGCP kitchen as particularly "bad"). What this excerpt says to me, then, is that while authority or expertise in the classroom can be limiting at times, it can also be used productively to encourage students' agency and creativity. The answer, it seems, is not to denounce agency or expert knowledge overall, but rather to learn how to use it constructively and strategically to legitimize students' varied locations and experiences, and to encourage students to forge new creative paths.

Most of the SGCPs teachers who I met and interviewed were interested in doing just this—in meeting the students where they were, and encouraging them to find their own agency within the food system. Unfortunately, however, there were also some leaders who were particularly uninterested in the food worlds of middle-school aged children, insisting instead that the only way to achieve effective results is by feeding students the foods, and food languages, that the adult experts deem appropriate. By not recognizing the students' food–body relationships as legitimate, and also rational and logical, the leaders end up reinforcing unhelpful dichotomies that pigeonhole (all) students as irresponsible eaters and alienate (deviant) students from experiencing the SGCPs in more positive and uplifting ways. The Edible Schoolyard leader Alice Waters, for example, insists that school gardens can teach students "what it means to be civilized," (a loaded claim to say the least) but her definition of what constitutes civilization is also predetermined by experts like herself (Waters Online). Thus, Waters' approach to SGCPs implies that what (some) students bring to the table—at least if it's fast food with preservatives, eaten sometimes, or pizza, eaten calmly on the soft couch—is not important to the success of the programs, and moreover, deemed to be uncivilized. This kind of expert positioning is condescending and alienating to students, as it denies the value of their own embodied experiences. But, as this chapter has illustrated, there are other ways to be an expert, or an adult leader, that allow for different outcomes to emerge.

Undoubtedly, there are many, contradictory ways in which both children and adults assert their agency as food decision-makers both within and beyond their SGCPs. The overall point that I want to make in this chapter is that despite the fact that the category of "youth" is often associated with a lack of ability to make food decisions, SGCP students are in many ways obvious and active participants in food decision-making. In some ways, the SGCPs acknowledged and encouraged this student engagement, while in other ways the SGCPs limited and undercut students' participation. In order to be successful at making SGCPs more relevant to students' lives, SGCP leaders and teachers would do well to recognize and attend to how, why, and where students assert their age-based food identities in an attempt to make food decisions "their own."

Part 3

Policy and practice

8   Food pedagogies

How might explicitly addressing body knowledge in education contribute to the philosophical sophistication of students' understanding of individual and collective agency?

(Springgay and Freedman 2007, xix)

The previous chapters of this book have explored how social differences impact students' experience of SGCPs through viscerally propelled mechanisms of knowledge production and identity work. This chapter moves forward from these analyses to address how the pedagogical strategies that the SGCPs employed both discourage and support diversity in students' visceral attachments to food. This discussion addresses the question of how bodies learn and develop within SGCP classrooms, giving particular attention to the degree or rigidity of the "scriptedness" of classroom activities. The focus on scriptedness follows from food scholar Mary Beth Pudup's important concern (2008) that tightly scripted classroom activities thwart students' agency within SGCPs, producing (potentially neoliberal) subjects that all think and act alike. Pudup's worry is that the disciplinary spaces of schools function as an instrument of "governmentality," in which students learn to self-discipline themselves in ways that reproduce a fixed and unprogressive structuring of food and bodies.

In examining and critiquing the pedagogical strategies and effects of the SGCPs in question, I therefore want to consider how the SGCP curriculum is (or is not) "bodied," following Stephanie Springgay and Debra Freedman (2007). In this chapter, a "bodied curriculum" is not just a statement that recognizes the body's presence within all forms of learning, although a bodied curriculum may begin with this recognition. Instead, it is an educational program that effectively bridges the gap between diverse, lived experience and abstract, intellectual work that is often so present in our academic experiences. Further, as the opening quote of the chapter suggests, it is a curriculum that consciously includes the body *in order to encourage students' agency* as individuals and collective beings. In essence, a bodied curriculum is opposite to "governmentality" or (neoliberal) "subject formation" (Pudup 2008). A bodied curriculum involves a type of learned, bodily discipline that does not limit but rather encourages diversity, and that fundamentally depends upon the engagement of all students as active, collective agents. A hands-on, sensory learning approach can therefore be part of a bodied curriculum, but it is what this approach sets in motion that is ultimately of consequence.

Below, I want to examine first how the pedagogical practices of the SGCPs are (or are not) involved in helping students to locate their (visceral) agency, and second how these (and other) SGCPs can potentially do this better. In my understanding, a bodied curriculum requires that educators use a conceptual framework that encourages students to understand their intellectual and bodily learning as an interconnected, developmental and embodied process—and also as a political practice. In such a program, students would be encouraged to locate their agency within the relational, but sometimes structured, flows that make up their perceptual experiences. They would be encouraged to find ways to experiment with visceral (re)action—to play around with not only the way they *think* about things like healthy, alternative food but also how they *feel* about such ideas and things, recognizing that it is the interconnection of both that drive how/who they are becoming. As I explore below, these practices require both discipline and flexibility, repetition and novelty, but they can also be constrained by such pedagogical forces.

Below, I take the discussion of pedagogical strategy beyond the oft-repeated sound bite that "it is the hands-on learning" (Teacher, CA) that makes SGCPs effective. In doing so, I ask not only *how* a hands-on program does its work (e.g. through repetition/scriptedness, or flexibility/unscriptedness, etc.), but also *what* work a hands-on program actually does (e.g. encouraging diversity, or homogeneity). Moreover, I want to stress the importance of both positive and negative visceral reactions in this educational process. I understand that for reasons of promoting SGCPs, "it is important to talk about the positives" (Leader, NS), but I also want to stress that it is crucial for the promotion of diversity to allow for and address the negatives. In this sense, it is important to keep in mind the frequency of the practice (in both SGCPs that I studied) of promoting a positive politics approach, in which negative associations or experiences of healthy, alternative food are eschewed in an attempt to promote positive experiences. There is an openness and potentiality to this practice, but as I have suggested in previous chapters, I think there is also a limiting, rigidness to it as well—a significant loss of opportunity. As Elspeth Probyn discusses of the importance of negative visceral (re)actions, "by bringing the dynamic of shame and disgust into prominence, we are forced to envision a more visceral and powerful corporeal politics" (Probyn 2000, 9).

Thus, in the sections that follow, I want to promote a navigated, fluid tension not only between scripts and spontaneity, rules and flexibility, or repetition and novelty, but also (and at the same time) between what we deem as a positive or negative experience: comfort and discomfort, excitement and anger, animation and antipathy. I want to suggest that not only are all of these experiences okay, and also somewhat inevitable, but they also can be ultimately productive to the process of understanding and promoting visceral diversity and bodily agency. In this sense, both positive and negative reactions should be acknowledged and discussed, even as SGCPs work more often to produce more of the former.

Rules and repetition

Rules and repetition in SGCP classrooms can have both "negative" and "positive" effects. I use negative and positive here not as judgments of a certain ethics, but rather as words that describe a relationship between bodies and the goals of the SGCPs; that is, effects that either turn students off to alternative food by limiting their control over food-based decision-making, or that encourage student involvement by making healthy, alternative eating more accessible and desirable to them. For example, students can become bored by the fixity of SGCP events, most notably when they feel as though their own role in producing food and food knowledge is limited by such fixity. Yet, as I witnessed during the course of my research, students can also come to appreciate the sometimes repetitive, perhaps ritualistic character of SGCPs. This occurs particularly as the routines of SGCPs allow students to develop new associations and familiarity with alternative food over the course of months and years of participation in the SGCPs. The routines and scripts of SGCP classrooms are therefore complex mechanisms of change that require balance and negotiation, but not outright dismissal.

Boredom

> Well, I have to tell you about the garden. In the garden we have a bunch of stuff that I can't name because I can't remember that much. Our garden is ruined in the winter. I'll tell you about the kitchen instead. I like the kitchen a bit, but sometimes I don't because we don't get to put food in the oven/stove.
> (Letter writing, NS)

The above statement is taken from a letter written by a SGCP student in Nova Scotia to a SGCP student in California. I begin with this student's lukewarm assessment because it captures well the energies of a student who feels somewhat "turned off" by the program. While not all students would describe their SGCP experiences with such halfhearted interest—this student's boredom is indeed almost palpable—in another sense many of the SGCP students articulated to me similar moments of tedium over the course of their fluctuating SGCP experiences. This is especially true of how many students described to me their experiences of the kitchen relative to the garden. While not many students were as dismissive of both the kitchen and garden experiences overall, most students in the SGCPs that I studied seemed to greatly prefer the kitchen activities to those of the garden. The reasons for this are multiple, but one explanation is that although both kitchen and garden involved some amount of rules and structure, the kitchen was more often a flexible space where students are able to negotiate their own role in its functioning (though perhaps not always in regard to the use of the oven, as the student above laments). Because gardens require a good amount of advanced planning, a teacher or adult supervisor would often pre-arrange the daily tasks to be undertaken in the gardens. By contrast, the kitchens were spaces where there were some rules (e.g. not touching the oven in the Nova Scotia SGCP), but also where there was quite a bit of negotiability in terms of what each student can do to contribute to its functioning. That being said, both of the garden programs that I observed also incorporated quite a bit of flexibility into the garden routines, as I will discuss in more depth in the section on flexibility and novelty below. In fact, relative to stories that I heard from other garden programs in Berkeley, in which "strict garden rules prevent[ed] any sort of negotiated experience between students and teacher" (Field notes, November 2007, CA), both Plainville and Central Schools embraced a good amount of unstructured activity in coordinating the students' garden experiences.

As Pudup (2008) claims, however, "the real [or most important] action is in the kitchen" (Pudup 2008, 1236). While gardening for most students does not represent a familiar activity of everyday life, cooking and eating are common activities that are positioned to have more of a direct impact on the daily life habits of individual students. Indeed, the imagined act of "taking the lessons [of SGCPs] home" (Teacher, CA) usually referred to taking a recipe or cooking idea home, rather than a method of raking or weeding. It is for this reason that Pudup (2008) particularly worries about the ability of students to find agency within a SGCP *kitchen* class. Scripted routines and repetitive acts in the kitchen are arguably more likely to influence the food-based decisions of students, because kitchen activities are, by their commonness, "closer to home."

While the kitchen spaces in Plainville and Central School's SGCPs tended to be less scripted than those of the garden, over the course of my fieldwork I certainly witnessed struggles between students and teachers (or other adult authority figures) that were related to the rules and rigidity of the kitchen classrooms. These struggles demonstrate that students sometimes feel as though the rules of the kitchen undermine their ability to assert themselves in cooking activities:

> We like to cook, but not with the teachers on our backs, like we are gonna burn something.
>
> (Student, CA)

> I cook all by myself, when there are no grownups. Then I would like to eat it . . . cooking is fun [here] but you always have to follow rules and stuff.
>
> (Student, NS)

The above comments suggest that the presence of adult authority figures, who enforce rules and structure cooking experiences, can detract from students' enjoyment of the SGCPs. Particularly, these comments reveal that students would be more inclined to enjoy cooking and eating (healthy) food if they did not feel pressure from adults to do so. While fixed rules and routines can add to this sense of pressure in the kitchen, the amount of rules and rigidity in the kitchen activities is also determined by particular personalities and moods of different teachers. Though there were indeed some "fixed" rules and standard routines in the SGCPs that I studied, such as "No running with knives," "Don't touch the oven," or "Each student must wash their own dish," a tightly controlled cooking experience was often reflective of a teacher's personal management style, which sometimes changed depending on the circumstances of a particular class. For example, in Central School, Ms. Tiffany tended to have a different approach to guiding her tables than another teacher, Ms. Lisa, which influenced their students' relative experiences of daily SGCP classroom events. Ms. Tiffany's morning table consisted of a group of eight students that tended to "slack off" and thus require more direction, according to Ms. Tiffany:

Field notes, October 2007, CA

Ms. Tiffany is pretty controlling with her morning group of kids. She came in late today, because she got caught in traffic, so Ms. Lisa had already said to the kids that she was going to take a back seat approach and let them practice their self-organizing skills. But then Ms. Tiffany came in and started micro-managing. Put this in this bowl, and that in that bowl. Some of the kids were obviously a little taken aback, and one girl actually said, "I thought that we were supposed to do it on our own." But Ms. Tiffany didn't hear or acknowledge this. Melissa was cooking the rice and one of the twins wanted to help, but Melissa sort of wanted to control it because the other girl was kind of flaky and kept on walking away and not really taking control. But then Ms. Tiffany got harsh with Melissa and said that she needed to share the responsibility. So then Melissa let the one twin stir for a while, but then the twin left again and she took it back over. Later there was a problem with the rice because someone hadn't added enough water, and so Ms. Tiffany had to add more and fix it. When we were sitting down to the table, Melissa sort of got blamed for this, even though I don't think it really fell on her.

This field note excerpt illustrates that strict control over the practices of a kitchen classroom can lead to tension between students and teachers, as well as hostility among the students themselves. Ms. Tiffany's style of "micro-management" in this example clashed heavily with Melissa's personality as a leader in the kitchen, making Melissa frustrated with both Ms. Tiffany and the other (less engaged) students. This sort of tight control can thwart students' abilities or desires to take control over the practices of the kitchen, making them less likely to want to engage with alternative food practices in the future. Furthermore, this type of rigidity reinforces the notion that the adults are the experts in the situation—the only ones capable of management—and that by contrast the students are there to learn from them (rather than *with* them). By so limiting students' role in the collective production of food-based knowledge, micro-management can thus limit their ability to locate their agency within such food events.

Beyond cooking practices in the kitchen classroom, rules about what specifically is good to eat also certainly factored into the scripts and routines of the SGCPs. For example, as previous chapters also discussed, many of the complaints voiced by parents in Nova Scotia were based upon the notion that the food itself was too strictly policed, that there was not enough variation (of junk and healthy food), and thus that their children would not eat it (a claim that was not always confirmed by the students' actions). In addition to parents, students also complained during their interviews about the rigidity that surrounds what is allowed to be cooked/served in the kitchen:

> I think that you can use guides to figure out stuff, but you also have to figure it out for yourself. I think you have to find your own way to eat. You can't always be healthy because it gets boring. Eating the same thing every day is boring.
>
> (Student, CA)

> The thing that now in schools, everything has to be healthy, that is going a bit overboard. Because I like junk food too. There should be a mix. How will we learn how to make choices?
>
> (Student, CA)

> There is no pepperoni anymore, just cheese. I mean, I am not complaining because I believe in healthy eating, but some of the stuff . . . [is over the top].
>
> (Student, NS)

These students' comments suggest that rigid nutritional guidelines are not only boring or off-putting but are ultimately unproductive in that they do not give students an opportunity to learn how to make their own decisions—to "find [their] own way to eat." Notably, the students do not necessarily or entirely reject the guidelines themselves, but rather reject the rigidity with which they seem to be enforced. I found that this was especially true in regard to lessons and claims about the healthfulness of particular foods. Students found health information to be important, and relevant, but they also stressed that it could be overbearing if health concerns structured every eating experience within SGCPs. One girl complained during a group interview: "You hear the gardening teachers and cooking teacher talking about how you need to eat healthy and you hear it all the time, everywhere, and I am really tired of it all. I just want to eat my food!"(Student, CA).

Similarly, during a peer-interview activity, another student also suggested that emphasizing health detracts from the otherwise positive experiences of cooking in SGCPs:

> *Peer interviewer*: Do you like the foods you learn to make in cooking class?
> *Peer interviewee*: Yeah, in cooking class, but not at home. It is harder at home without my friends' energy around. Some things are good, definitely the bruschetta. The pancakes were a disaster . . . a disgrace! I think the healthy outlook ruins the food, especially if they make it an obvious point. I like the cooking, but also the atmosphere. But some things don't work at all.
>
> (Students, CA)

In the above excerpt, a student discusses her reaction to SGCP foods. It is obvious that she enjoys many aspects of SGCP foods, and especially being in class with her friends. However, she says adamantly that, "the healthy outlook ruins the food, especially if they make it an obvious point." Her comments imply that healthy eating *as a rule-based activity* is unappealing. She would rather not know about a food's health status, and would rather not practice food activities in terms of following such guidelines or rules. In this sense, repetition and rigidity in terms of not only the food itself but also how the food is discussed is an important factor in how students come to experience SGCPs as (potentially) overbearing and strict. Certainly there are many ways in which rules and repetition in SGCPs can limit students' visceral access to alternative food.

Familiarity

In other ways, however, rules and repetition in SGCPs can be productive in allowing for new experiences of healthy, alternative food, especially by encouraging students to become familiar with particular foods or food practices. In this sense, the familiarity and comfort that come from repeated actions can allow students to relate to, and possibly even come to enjoy, healthy alternatives that they otherwise may not have had a chance to experience. Indeed, many of the SGCP teachers with whom I spoke recognized that the rituals, scripts, and repeated actions/recipes of their SGCPs were an important mechanism through which their particularly disenfranchised students—those with less "access" to alternative food, for example— became more accustomed to new foods and food spaces.

In garden activities, SGCP teachers commented that repeated labor in a particular area of the garden seemed to encourage feelings of responsibility and ownership among many students. In fact, during my research tenure, several garden teachers/ volunteers in Central School were in the process of trying to brainstorm ways to further a sense of stewardship or ownership through such repeated association. In giving each class a space in which they could labor and make decisions, the teachers hypothesized that students would come to feel more personally connected to the garden: "I think that I would really like to have more distinct spaces in the garden where they can hang out, so they feel like they have their own special space . . . you can build ownership that way" (Teacher, CA).

Other teachers similarly commented that not only is repetition of work within certain spaces important, but that repetition of particular bodily actions can also be helpful in "reaching" or positively impacting some students. The repetition of particular actions like raking or digging served as a sort of "guided meditation" according to some teachers, which allowed students to make use of the garden space in a way that furthered their own emotional and physical well-being:

> I have heard of counselors who do walking or gardening counseling and I think that's really good. I think some of these kids have never learned to control their bodies and that allows them to calm down and focus. [One girl] got herself suspended for using her arms inappropriately. So I think this can make

her aware of her body ... And even the other girls who complain about gardening, I notice the conversations, once I get them working; by having something to be focused on physically they are able to reflect and have conversations that are important to their development ... So, I work on that personal development level and support and self-empowerment and how to use your body and be focused.

(Teacher, CA)

In the kitchen classrooms of SGCPs, a similar sort of effect can be witnessed in regard to the function of repeated actions and routines. The routines of the SGCP kitchens that I am referring to here include everything from the standard ways that chores were divided up and carried out within the programs to how the act of eating itself was repeatedly conducted. Some teachers even called such routines "rituals," reflecting that there is a sense of greater purpose and higher cultural meaning in the act of engaging in routine:

We have a routine here where the students go around the table and say what we are thankful for, prior to eating in cooking class. I would think that most of us would agree that it is healthy for families to eat together and share chores together, and use that as a chance to communicate, and so they are getting that in a school setting, even though they might not be getting it at home.

(Teacher, CA)

As the above passage suggests, in the Berkeley SGCP, students repeatedly ate collectively around a table, often sharing with each other reflections on their day as they ate together. This collective eating was an activity that students quickly came to expect as the finale to their daily cooking routine. (Students in the Nova Scotia site also ate together following the cooking activities of the day, in a less formal though still anticipated manner). In regard to the act of gathering collectively at a (set) table, many of the student participants in my research found this routine to be comforting, even if they did not engage in such an act during their food-based practices at home. In fact, many students commented that eating collectively around the table was something that they were particularly fond of, both for the opportunities it afforded to socialize with friends, and for the time it allowed them to withdraw from a busy school day to relax and eat a snack. This is particularly noteworthy because, as I discussed in previous chapters, the ritual of a shared meal around a table also signifies a sort of moral correctness for some SGCP leaders, which could lend itself to the production of sensations of judgment or resentment among students who do not eat in such a manner at home. However, for the most part, the repeated act of eating together at a table was something that—even in contrast to other modes of eating—came to be expected and enjoyed by many students.

Like the routine of a shared meal, the act of collaborative cooking (or even cooking at all) was also new to many students. While some students were therefore hesitant or unsure when first entering a SGCP kitchen, the incorporation of cooking into the daily routine of SGCP schools allowed the activities of the kitchen to become familiar to these students. As students became used to the smells, sounds, and general bustle of the cooking class, cooking itself became something that was more accessible to them. In other words, with repetition, even cooking itself became less strange, and thus more fun. A student in Nova Scotia wrote to one in California, explaining: "I've never cooked for the whole school before in my life. I only cooked for my friends and family. If you cook long enough it won't be weird, it will be fun" (Student, NS).

As the above student simply suggests, after cooking for a while, the act of cooking switches from feeling strange to fun. The same can also be said for particular items of food that are strange or unfamiliar to students. In addition to the act of cooking, teachers, parents, and students often repeated a similar story about the SGCP food itself: that while the initial reaction of students might be, "This is bad," or "I don't know this food" (Teacher, NS), by having the food in the kitchen repeatedly, "it becomes less foreign" (Teacher, NS). Indeed, this particular function of repetition was one of the reasons why so many SGCP teachers and leaders suggested that it was important to "get to the kids early" (Teacher, CA), as I discussed in the previous chapter. After years of relating to a particular food, the teachers observed, the food becomes more accessible to students:

> Openness to new foods is much bigger now than it was before, because many kids have been cooking since kindergarten. Brown rice used to be foreign and tofu, no one wanted to touch it. Now it's not that foreign, and from the beginning to the end of year, the kids didn't like a lot [at the start] and by the end of the year you saw a progression. So there is more awareness around food.
>
> (Teacher, CA)

> I think kids are more open this year. I think that's testament to kids having been through programs at elementary schools [in Berkeley]. We see the effects. And it's not just the Berkeley hippie parents, it's kids from all different classes. I saw two African American girls making fun of another for not knowing the difference between a cucumber and a zucchini.
>
> (Teacher, CA)

While these points regarding repetition and learning may seem obvious, it is important to point out that there is more than one way to interpret this process of what we might call "normalization." While on the one hand the work of repetition could be seen as a homogenizing force (in that it encourages the development of familiarity with similar tastes) it could also alternatively be read as a developmental process that expands rather than precludes visceral difference. Indeed, in the SGCPs that I observed, the latter scenario is more likely to be an accurate reflection of the work of repetition, especially because over years of engagement, it appears that most of the SGCP students that I spoke with have effectively negotiated particularized relationships to SGCP food and food practices that would not be considered especially homogenous. For example, certain students look forward to particular recipes, while others may enjoy working with a particular area of the garden, or within a particular part of the kitchen. That is to say, the impact of repetition overall, it seems, has not simply been one of increased homogeneity. Moreover, some students may still hate eating zucchini or cucumbers, even if the *idea* of eating these foods is no longer new. I witnessed all of these scenarios, as well as others, while working within the SGCPs.

A final way in which repetition is important to the functioning of SGCPs is through the actions of alternative food activism more broadly. In this sense, I use repetition to signify the act of continually "sticking with it" despite hardships or setbacks. In other words, the processes through which SGCPs come into being is often long and hard, perhaps involving criticism from the local community, lack of ability to garner funds, or simply a never-ending time commitment in terms of keeping the program afloat. In such scenarios, repeating one's actions, even to the point of rigidity and inflexibility, is sometimes a necessary means to an end:

> I just kept my head down and kept doing it, the only way to make change is to keep going with it for some years. So I didn't go out to find out what people were saying.
>
> (Leader, NS)

> Alice [Waters] really believes in [her vision] and there is something very powerful about that. You have to have that intensity and rigidity to get stuff done. So that's cool, nothing wrong with that except she's a little Francophile and that's not appealing to everyone.
>
> (Activist, CA)

In these statements, the activist/leaders suggest that some amount of rigidity is necessary for strategic reasons in order to get the SGCP initiatives off the ground, and also to keep them going. While it is also certainly important to examine the ways in which rigidity in activism precludes certain individuals from acting at all, these activists' statements suggest that flexibility and fluidity are not the only (potentially progressive) modes of operating. Indeed, much like the role of repetition in regard to the familiarization of food itself, my research also revealed that the very act of repeatedly running a SGCP within a local community, year after year, is also a routine that is crucial to the functioning of the SGCPs. In fact, many leaders and teachers from both programs simply credit the repeated presence of the program itself as a central component of their SGCP's overall success, by slowly but surely familiarizing parents with alternative food, and by eventually even convincing some that the programs are worth supporting. In this way, many parents who were at first skeptical of the programs came eventually to view SGCPs as beneficial (or at least benign) over months or years of relating to these programs.

Flexibility and novelty

Like rules and repetition, flexibility and novelty within SGCP classrooms can have both "negative" and "positive" effects, either by making students frustrated and uncomfortable with alternative food, or by making such food more pleasing and exciting. Students can become overwhelmed by chaos within SGCP events, and can come to feel directionless in regard to their own role in the programs. Yet students can also thrive within conditions of flexibility and novelty, especially because such characteristics lend an openness to SGCPs that is conducive to encouraging active input from students. Thus, the flexibility and novelty of SGCP classrooms are, too, complex mechanisms of change that require balance and negotiation with pedagogical practices of repetition and rigidity.

Discomfort

To be sure, this research has stressed the importance of flexibility and fluidity in both theoretical and practical terms. This emphasis is important for many reasons. Yet, insisting on flexibility always and everywhere can certainly be counter-productive. There are moments within SGCPs when flexible, unstructured activity can become too much, meaning that it can produce sensations of discomfort or unease among some students that detracts from their ability to engage with alternative food or locate their agency within SGCP activities. Of course, as I have argued in previous chapters, it is not possible to fully predict when or for whom these negative sensations will arise. Yet, even so, it is important to try to understand how flexibility in SGCPs can lead to such negative reactions, so that educators can more effectively negotiate a balance between structured and unstructured events.

In regard to the garden work of SGCPs, my fieldwork reveals that while many students cherish some amount of unstructured play, unstructured activities that go on for too long, or are very chaotic, often produce sensations of boredom or frustration among students. Students sometimes view the garden as a place to relax and unwind from the other more structured activities of their school day, and yet at times the leisurely feel of the garden seems to produce less favorable experiences. In writing a letter to a student in Nova Scotia, a Berkeley student reveals: "Well, besides saying that my school garden and cooking class is cool it is also a way to relax . . . At times I think it can be boring" (Student, CA).

To the above student, the space of the garden is both relaxing and also potentially boring—a combination that suggests, perhaps, a need for a more coherent aim. In my discussions with teachers about the frequency of student complaints about gardening, I came to interpret the boredom or frustration that some students expressed as stemming from a lack of direction or purpose in the garden. In this interpretation, too much flexibility was limiting to students' ability to access positive visceral reactions in the garden, particularly because in the (often) unfamiliar space of the garden, the students required more guidance as to how to approach their garden work. In other words, some students (at least) seemed to be at a loss as to how or where to fit themselves into the garden activities, and thus tended to resort to simply following (boring) orders; these students needed help "accessing" the garden in a different way.

In addition to teachers, I questioned SGCP students themselves about their relationship to the garden, particularly in terms of their sense of "ownership" (a oft-repeated SGCP sound bite). I often asked whether or not the students felt as though the garden was "theirs." Although some students suggested to me that they did feel as though they had some claims to the garden, having worked in it, few said that they felt connected to it in the sense of being active agents in the reproduction of the garden space. In other words, few students felt like they were in control of the garden, or that they knew the garden well enough to take a leadership role in garden activities. I came to understand this lack of connection as central to the students' disinterest in the garden (relative to the kitchen), and as stemming at least partially from an inability to sense their own positionality within the broader picture of the garden as a whole. Garden activities seemed haphazard and pointless to students, because the larger functional plan was inaccessible to them. Consider the following dialogue that I had with one student (during a group interview) at Central School:

> *JHC*: So, you don't like the garden part of class, then?
> *Student*: Yeah gardening is weak because . . . it's weak. I didn't sign up for no garden class.
>
> (Student, CA)

> *JHC*: OK, so what do you think would make it better?
> *Student*: Um, like. I don't know. Just make it better, and stop wasting time. We waste too much time. I mean it ain't on you all, it's on us [too] cause we don't pay attention and be talking and stuff, [so] like less of that.
>
> (Student, CA)

As the quotation above suggests, this student becomes frustrated with the garden class particularly when she feels as though the teachers are just "wasting time." As her comments imply, more structure (and potentially more rules—e.g. requiring students to pay attention and refrain from talking) would probably help the garden class run more smoothly. My interviews with some SGCP teachers also revealed a similar interpretation:

> I think one thing missing from the garden is a purpose. Sometimes they don't understand why they are there. And we can give them a small skill idea like forking, but sometimes they aren't even planting the next week and sometimes the bigger sense of why that space is important and why they should feel good about working in it isn't really there.
>
> (Teacher, CA)

In this teacher's assessment, the students feel disconnected to the garden when they cannot sense a greater purpose for their being there. Activities in the garden are not structured or progression-oriented enough to allow students a sense of consistency and coherence in what they are doing, and thus the lessons feel overwhelmingly haphazard at times. I also witnessed a similar dynamic in the SGCP cooking classes. Although students generally tended to enjoy the cooking classes better, the SGCP kitchens also had the potential to become frustratingly haphazard spaces. The event with a professional chef in Plainville School, for example, seemed to border on uncomfortable chaos for some students. Even though many found the process thoroughly entertaining, the rushed and hectic event was too unstructured for students to do much else than follow the chef's orders (which ironically reinforced a sense of rigidity to the event). In this sense, some students did not find the event particularly engaging or relevant to their own lives.

Most of the SGCP cooking teachers that I spoke with, however, seemed to recognize the importance of at least some amount of routine within the processes of cooking and eating food, and thus tended to incorporate a balance of structure and flexibility into their lessons. As a cooking teacher from a related school within the Berkeley school system suggested to me, too much flexibility can indeed be overwhelming to students, while a routine can make cooking seem more accessible (and therefore, more fun):

> It is really important to establish a sense of rituals, and have some structure because *it is overwhelming when there is no structure*. I divide class up into a very clear beginning, middle, and end, with distinct tasks like cutting, or boiling, or setting the table. It doesn't [seem like a lot of work] that way, it seems do-able and fun.
>
> (Teacher, CA, emphasis added)

Beyond cooking and gardening activities, flexibility in regard to taste itself can also sometimes become counterproductive to the goals of SGCPs. That is, insistence on a flexible *approach* to eating can sometimes produce feelings of discomfort or distaste among students and parents alike. As other chapters have discussed, when SGCPs ignore the established food habits and routines of certain students, families, or communities, reactions to SGCP food can be negative. We saw this dynamic, for example, when the Plainville chef produced exciting, novel tastes for the "unsophisticated palates" of parents and students within the school's local community:

> Honestly, I didn't really like the taste. It was a little weird for me.
>
> (Student, NS)

> I think the things he made were just a bit too different from what people are used to. The things they make with Ms. Dora are more recognizable to the parents. It is a balancing act.
>
> (Teacher, NS)

Although many SGCP teachers obviously find it important to encourage students to be increasingly open to new tastes, as the above quotes suggest, in order to reach segments of the population that do not already have a way to viscerally "access" alternative food or flexible eating in general, it is also necessary for SGCPs to pay attention to the already established routines and rituals of the broader community. Indeed, particularly "rigid" or solidified food habits, preferences, and traditions need to be acknowledged and negotiated if SGCP initiatives are to encourage increased participation and increased "openness" to alternative food from a more diverse population. Moreover, as previous chapters explored in regard to class-based food hierarchies, the demand for flexibility or openness in regard to taste can itself become an implied "should" of eating within SGCPs. Indeed, flexibility can come to describe not inclusive eaters, per se, but rather cosmopolitan eaters with a wide array of elite food habits that are nevertheless usually *not* open to foods deemed unsophisticated, processed, or unhealthy. Along these lines, inflexibleness to food then comes to signify the food preferences of unrefined, "white bread" populations with a limited food vocabulary. Surely these associations are not productive for the promotion of widespread openness to alternative food. In attempting to encourage all students' agency within SGCPs, then, it is important to recognize that inflexible eating habits are also legitimate ways of knowing food alongside openness and flexibility. Indeed, in regard to taste, "it is a balancing act" between novelty and routine.

Comfort

While flexibility within SGCPs can sometimes create problems, my research experiences revealed that more often than not, a flexible approach to food-based learning invites students to negotiate their own identities and positionalities vis-à-vis alternative food, and thus allows students to move forward in the task of making healthy eating "their own." In regard to the pedagogical strategies of teachers, flexibility in this sense refers to anything from allowing for informal conversation among peers to welcoming spontaneity in classroom decision-making to including students' desires or interests within the broader educational process of planning and structuring the SGCP. Often, but not always, such flexibility was more readily found in the SGCP kitchens than in the gardens:

> I feel more like it's my kitchen when I am in the kitchen then in the garden, because you are assigned what you do [in the garden]. In the kitchen, you can pick. And they are a lot more flexible. You can use the knives when you want. They aren't like don't do this and that. And in the kitchen you feel like you are contributing more because you eat the food. The garden feels like it stays the same the whole way through but in the kitchen you make a good meal and eat it.
>
> (Student, CA)

As I have suggested, in both of the SGCPs that I studied, the kitchen space was frequently deemed the more "fun" space to be. Usually this was because the kitchen was where students were encouraged to choose what part of the meal they wanted to help with, which spices they should add or leave out, or how they would set the table, among other such activities. Flexibility as an understood but unstated practice in the kitchen seemed to encourage students to take control of the functioning of the kitchen space in a variety of different ways, perhaps partially because a flexible approach on the part of the teachers illustrated to the students that they were both trusted and counted upon to take a leadership role. In contrast, as I have discussed, the garden was a space where there was less room for flexibility, both because the options for picking tasks were more limited, and because the students themselves were less sure what sort of tasks were available as options to choose from. In this sense, student-led activities in the garden were hard to come by.

In some ways, however, even the garden space was experienced as a flexible space. For one, in contrast to other standard classrooms within both SGCP schools, the garden represented a respite from the normal(ly more rigid) structure of the school day. In addition, the emphasis on hands-on learning, as opposed to intellectual or book learning, provided different sorts of opportunities for more flexible educational experiences. As one classroom teacher described: "I think it gives a chance for kids to experience a space that can be theirs and figure out how to move in that space and . . . I just think it gives them a chance to get out of a classroom setting" (Teacher, NS).

In addition, the SGCP garden teachers also acknowledged that they benefited especially from the opportunity to interact with students in relatively small groups. The informal feel of the garden classroom, combined with the closeness of the bond between SGCP teachers and students, as well as the opportunities in SGCP classes for intimate and free-flowing communication, allowed for experiences that could not have come about in a standard, structured classroom:

> We can have closer bonds with the kids, different bonds than a classroom teacher. We have smaller groups, and we can hang out with them more. If we know about their own lives, chat with them, they are more likely to listen. We have that luxury.
>
> (Teacher, CA)

Indeed, in my experiences in the SGCPs, the garden spaces provided students with opportunities not only for open communication with teachers, but also free-form interaction with the land and with other students. For this reason, the garden teachers often deemed the end result of a garden activity as less important than the collective, haphazard process of working the land together. In fact, some of the garden teachers approached their teaching in a purposefully flexible way, inviting daily activities to unfold as they may, rather than sticking insistently to a preplanned objective:

Field notes, December 2007, CA

Liz had an interesting group today. She got them together to smooth out the garden bed, which was sort of tedious for a lot of the kids (who were visibly looking tired and annoyed). But, then they planted cover crop, and they really seemed to like this. Liz had the seeds in a bowl and explained that they will help the soil to have nitrogen in it, for nutrients. One girl said the seeds looked pretty, like African rice. Another boy asked if he could taste it (and Liz let him). All the kids really wanted to get their hands into the seeds, to really plunge them deep into the bowl. Latisha looked at me and said she wants to have a spa when she gets older, where people can put their hands into bowls filled with seeds. (She usually hates gardening.) We went to the beds, and poked small holes all over the top, putting three seeds in each hole and pinching them closed. One of the boys said, "Pinch, pinch, pinch," in a weird voice over and over again, and told me that he was tucking the seeds in. Liz told them to sprinkle some more on top, and then each make a wish.

In the description above of a particularly successful garden activity, the garden teacher, Liz, employs some amount of structure in the format for the class (e.g. she had some particular goals in mind—smoothing beds and planting top soil). Nevertheless, it is the unstructured events of the day that quickly became most important to the garden experience. For example, all of the students want to plunge their hands deep into the seed bowl, in order to feel the interesting sensation that the seeds provide. Liz welcomes and heeds this request, as well as the request to taste a seed, and the result of this welcoming is not insignificant: Latisha, who normally hates gardening, actually has an enjoyable and memorable experience in the garden. So, too, do the other students, who all enjoy planting the cover crop. In this sense, by allowing and encouraging unplanned events, Liz allows the students to connect to the garden in their own way, giving them a sense of interest or even commitment to the space (e.g. the "tucking in" of each seed).

During my fieldwork, SGCP kitchen teachers often took a similar approach to the kitchen classroom. Being flexible in the kitchen was certainly a way for the cooking teachers to remain sane in the midst of potential chaos (as many of us can imagine, the task of teaching multiple middle-school age children to cook is not easy), but it was also a way for the cooking teachers to convince the students that cooking could be fun. In my observations, if the cooking teachers allowed students to turn on the radio, talk about a dispute with a friend, or try to make a recipe that was different than expected, the result of such allowances was more often than not a positive experience for all. Importantly, for several of the cooking teachers, this flexibility included talking only minimally or indirectly about the healthfulness of food or the supposed "shoulds" of healthy eating:

I have to be very relaxed. That is why it works. I don't talk about healthy eating outright. I just show them that cooking can be fun. It gets them thinking about food a bit more.

(Teacher, CA)

They can talk about anything in here, and they are told not to repeat it outside. I mean, if something is very serious I would report it to the principal, but generally speaking this is an open forum to talk about whatever is on their mind. And they have a lot of fun, they can be relaxed . . . I am pretty laid back, you have to be to do this.

(Teacher, NS)

I don't know if I teach them, I don't really. They just learn by doing.

(Teacher, CA)

As the above comments reveal, in these cooking classrooms, there is very little, if any, formal instruction. Indeed, part of the appeal of SGCP classes for many students is the break in the day from normal classroom activities. In addition, the kitchen spaces themselves are, much more than other classrooms at the school, often also used as informal "hang out" spaces for the schools' teachers and staff (as well as for straggling students who are late for other classes). Most likely, this "hang out" function develops over time as the food cooked in the kitchen attracts the attention of passers-by, who come in to the room to see what is going on. This scenario was true especially in Plainville School, where the SGCP kitchen was also the place from which school lunches were served. The administrative staff would therefore often drop by to chat with Ms. Dora or the students, and to get a taste of the food. This informal and unplanned use of the kitchen space demonstrated to the SGCP kids that the kitchen is a fun place to be.

Both SGCPs that I studied also had special events in which flexibility was encouraged in certain ways above and beyond what tends to occur in the everyday activities of the SGCP kitchens. As I described in Chapters 4 and 6, at Plainville School this event was the once-a-year cooking engagement with a professional chef, where the recipe was set in advance but the kitchen activities were experienced as hectic and disordered. In Central School, the event was a monthly cooking competition in which the teachers took a back seat approach and gave students (almost) full reign of the kitchen space. In both of these events (albeit in different ways), students were invited to be especially creative with their cooking endeavors, and teachers remained flexible to respond to the students' interests and whims. The result, in both cases, was contained chaos: "They all love it. It is craziness in the kitchen, wild" (Leader, NS).

Beyond wild and crazy cooking, flexibility in the kitchen also comes to be important to the success of SGCPs in terms of actual tastes or food preferences. Flexibility in the sense of taste has two meanings. First, I found that SGCP teachers generally have more success with convincing students to try something new when they are flexible in their demands. In other words, flexibility as a practice in persuasion seems to be important in motivating students to taste new foods. One teacher told me particularly that she "was trying to meet the kids closer to where they were coming from," because doing so made it easier for students who were unfamiliar with alternative foods to relate to the program. Another SGCP garden teacher explained that she invites students to sample food from the garden without making it a requirement: "I think . . . [the idea of] exposing kids without pressure is important. . .having that free access to something enables kids to be like huh, what is that? . . . I find that by the end of the year you can get a kid to taste something" (Teacher, CA).

The second meaning of flexibility in regard to taste has to do with the novelty of the food experiences themselves—the wildness or craziness of the tastes. While I have already expressed the need to pay attention to pickiness and rigid habits when it comes to tasting foods, it is equally important to allow for and encourage variation and experimentation when it is feasible to do so. As several teachers expressed to me, allowing for variation in the recipes and routines of the cooking class is often a way for them to promote interest among students and attract their attention or peak their curiosity in regard to alternative food. Unusual events or opportunities can also signal to students that the SGCP initiatives as a whole are unique, and that what the SGCP offers to students is valuable and significant. In this sense, the novelty of unique foods and food events also can provide students access to sensations of intrigue in regard to alternative food:

> Doing something unusual is important, like the BBQ in the garden, it is unusual and it is free, and you don't make them do it, you offer it. And they all want it.
>
> (Teacher, CA)

> I like it cause you come here and get something really different every day. Most of the stuff we cook here I never ate. And the stuff is healthier than the stuff we eat.
>
> (Student, CA)

In all of the above ways, flexibility in the SGCP classrooms encouraged students to become actively engaged in cooking and gardening activities, and allowed them to more effectively negotiate their own particular relationships with alternative food in ways that matter to them. Undoubtedly, some amount of unstructured activity—balanced by a repetition or routine—is therefore crucial to the successful functioning of SGCPs in regard to their ability to motivate students to eat "healthier," alternative foods.

Concluding thoughts

> I like to give the kids more freedom. It is a challenge having the kids use knives and so on. We negotiate and try it out, but its fun. The kids have a wonderful time. It's a kid-driven program in my mind.
>
> (Teacher, NS)

> Creativity seems to emerge during the iron chef event, how they set the tables, come up with plates for judges . . . a table of boys did some amazing things with a mango, cut it up in the shape of a pineapple, [it was] very artsy.
>
> (Teacher, CA)

In this chapter, I have drawn on data from interviews, participant observation, and in-class activities in order to explore how the pedagogical practices of SGCPs both limit and reinforce students' agency in regard to food-based decision-making, within and beyond the classroom. In particular, I examined the effectiveness of two broad pedagogical strategies in the promotion of visceral diversity and bodied agency: rules/repetition and flexibility/novelty. And, I described how each strategy both constrained and promoted SGCP students' access to healthy, alternative food as something that is (relationally) their own. Returning to the discussion that opened this chapter, we can see how the above experiences speak to Pudup's (2008) important concern that SGCPs can promote rigid subject formation, but also to Stephanie Springgay's and Debra Freedman's (2007) vision for a more contextual and negotiated "bodied curriculum." While neither SGCP *wholly* succeeded at helping students to bridge their diverse lived experiences with the intellectual pursuits of their SGCP classrooms, it is clear that the "hands-on" focus of these classrooms does allow for spaces where such bridges can be assembled. As their experiences of creativity and alteration suggest, SGCP students undoubtedly do more in SGCP classrooms than embody the (*potentially* neoliberal) rules of leaders and advocates—even if they do also do this (Hayes-Conroy 2010a). Furthermore, the teachers with whom I worked during the months of my fieldwork were all well aware of the complexities and contradictions of what they did in the classroom. And, they were tremendously adept at navigating this complexity in order to encourage students' agency as individuals and collective beings.

Conclusion

A thousand tiny eithers; a thousand tiny ors

Judgment is really toxic when it comes to food, and I think that the . . . poo-pooing cheetos and twinkies and so on, what that is doing [to] children is it is creating these complexes where children feel bad about themselves when they want to eat those foods. I think there [needs to be] room for everything, and *room to listen in*. Food is cultural and emotional and social and nutritional and we need a space for it to be all those things.

(Activist, CA, emphasis added)

[Our school system] is one of the few places where [empowerment] has to happen . . . because that is where our kids are most of the time. That is where I was most of the time, and *I wasn't empowered*. I look back and think, gosh man, where would I be if I actually was given meaningful information, information that would really strengthen me and my esteem.

(Activist, CA, emphasis added)

In this book, I have explored School Garden and Cooking Programs (SGCPs) as instruments of both nutrition education and the alternative food movement. I have approached SGCPs particularly as viscerally driven programs, meaning that I have chosen to focus on how and why these programs physically motivate students to make changes in their eating habits, also addressing when and how they do not do this. In order to accomplish this analysis of SGCPs, I have drawn attention to the role that food knowledge(s), food identities, and food pedagogies play in the production of visceral "access" to alternative food. I have attempted to allow for explanations that draw attention to structural, epistemological and material forces, and I have insisted on paying attention to both how SGCPs reinforce and resist social norms, categories, and hierarchies. In doing all of this, I have described in detail how two particular SGCPs, one in Nova Scotia and one in California, come to be interpreted and experienced in a variety of different ways by students, parents, teachers, and food activists.

In what ways is such an analysis unique to SGCPs? Why do SGCPs matter especially in exploring such mechanisms of viscerally driven social change? In one sense, the answer to these questions is simple; SGCPs are not particularly distinctive to the visceral analysis that I have developed and conducted. What I describe in this book is not an exceptional phenomenon in the catch basin of worldly events. No, the point is in fact quite the opposite. What I describe in this book is true to life in many places; it is not unique but rather commonplace. Race, class, gender, and age are everywhere present as aspects of identity in our society, and they are always significant as axes of difference (among many others) that structure how we come to know and relate to the world around us. Certainly these distinctions are by no means particular to SGCPs, nor are they unique to the alternative food movement at large. Furthermore, in *most* social events or phenomena there will be instances of both reproduction of and resistance to these structured-yet-relational differences—for life itself both contains and contradicts such groupings. In justly describing any particular visceral existence, all of this is indeed present.

So why look at SGCPs as places in which these dynamics unfold? My point is not to claim that SGCPs are unique to such an analysis but rather to suggest that we must begin to allow such understandings of social life to inform our political practice *in a variety of places*, both within and beyond alternative food. To return to a point that I made in Chapter 2, as researchers we must learn to develop explanations that are always and everywhere infused with a "fidelity to what they describe" (Latham 2003)—a fidelity that allows for messiness and contradiction, incompleteness and potentiality, structure and agency. Indeed, as I have argued in this book, it is this sort of approach to research that can recognize and account for social difference in its full, visceral complexity. But we must learn to do this not just for the purpose of accuracy in our research, for this fidelity is needed beyond our academic scholarship. We need to begin to appreciate what it takes to put such an understanding into practice in activism, education, and policy-making.

As an example of the importance of this task, SGCPs draw our attention to how such a viscerally oriented analytical approach can become important to the promotion of progressive social change in social organizing, in curriculum development, and in health promotion and protection. My contention in this book has never been that SGCPs are the only place where such activities are important, but rather that the focus on "alternative" nutrition education within SGCPs is a useful place to begin to figure out how best to think through and indeed work through such complexities of everyday life. At least, the focus on SGCPs has been helpful to me in accomplishing this task. I began this research through my academic and personal interests in the alternative food movement, insisting that alternative food activists were not doing enough to reach out beyond lines of social difference (Guthman 2008a). I wanted to explore both why the alternative food movement has been arguably unsuccessful at attracting a diverse base of supporters, and also how movement leaders could begin to reach out more successfully. At the same time, I was attracted to this project because I was concerned about the way that education more broadly is being conducted in North American schools. I saw that the purpose and means of education were becoming increasingly limited and evermore conservative (Strober 2003), and I worried about the implications of these trends for the promotion of knowledge(s) as a collective and negotiated process. Finally, my interest in bodily health and nutrition *as an inclusive practice* led me to question the persistence of distinct "shoulds" in regard to ways of eating, as well as to question the expert mentality that comes alongside such health claims. I was therefore interested in searching for other means of attending to nutritional health in ways that were less judgmental and more engaging.

For me, then, SGCPs became a way to better understand how progressive social change can happen in the midst of social difference, and particularly, how the material body itself can become an instrument of such change. I saw these two projects as linked because, as I came to recognize, it is in and through bodily experiences of life—through our visceral (re)actions—that social differences are both reinforced and resisted. In this sense, by paying attention to bodily motivations to eat, I learned how we, as activists, educators, and health care professionals, might better "listen in" to a variety of personal stories and attachments (as the first activist describes in the quotations that begin this concluding chapter). In doing so, I was able to come closer to understanding and specifying how we might begin to move towards the elusive but nevertheless profound goal of individual and collective "empowerment" (as the second activist encourages in the quotation above).

In concluding, I want to now turn to briefly examine the lessons of this SGCP research for social activism, public education, and nutrition policy. In light of my body-centered framework, I will guidepost each of the discussions below by utilizing three central axes of analysis: the production of knowledge, the structure of power/ hierarchy, and the unstructured ontology of lived experience/social practice, keeping in mind that it is at the intersection of these three axes that everyday life unfolds.

Social activism and visceral difference

Feelings are important to activism. Many have argued this point (Thrift 2005, Hayes-Conroy 2009, Hayes-Conroy and Martin 2010, Clough 2012), and many more have experienced it as true. Indeed, in regard to SGCPs, motivation is what makes social change happen. As this book has argued, visceral reactions drive students to make judgments or choices about food, some of which lead them towards alternative ways of eating, and some of which steer them away from such foods and practices. There are many reasons for this, and many potential outcomes, but the first point to make about social activism and visceral difference is simply that bodies matter. In connection to and alongside the stated beliefs and ideals of any social movement, bodily sensations and experiences "move" us to act and (re)act. That our visceral reactions are not predetermined or fully predictable is central to this movement as well, for it is the haphazardness of our interactions that allows for new possibilities to emerge. All of these points serve to describe social activism as it unfolds within the unstructured ontology of lived experience. The stories from SGCPs remind us that activism in daily life necessarily takes place through complex, contradictory and chaotic networks of relation. To know and practice this ontological claim as a leader or organizer of social action is to recognize and encourage difference in knowledge, social position, and experience within a social movement.

Second, in regard to the structure of power/hierarchy, this SGCP research illustrates clearly that visceral reactions are not natural, pre-political bodily impulses. On the contrary, visceral reactions are always already imbued with a certain politics—a developed array of tendencies (yet also latencies) that tell broader stories about the inequities of access that color our social world. Such stories include a wide array of structural injustices—in this research, for example, discrepancies in geographic placement of food markets in different (often racially and economically defined) neighborhoods, or disparity in purchasing power among different families. Such stories can also highlight variations in cultural capital, education level, or availability of social services, or they can give us information about the social norms and expectations that confront a variety of individuals and groups. To a social activist, then, recognizing visceral reactions as socio-political is an important step towards understanding how to respect and reach out to groups of people who are "other" to the social movement or organization at work. This is a particularly important point for the alternative food movement, which has been critiqued for its homogeneity, yet it is also a critical lesson for many other social movements. By beginning to recognize how our bodies have developed differentially in and through the social hierarchies we wish to dismantle, we can better learn how to resist those hierarchies through developing new and accessible opportunities for interaction.

Third, this SGCP research points to the need to question and dismantle expert knowledge claims within "progressive" social activism. In insisting on the hybridity of mind and body, we can begin to appreciate how knowledge itself comes to matter materially in the production of bodily motivations. Knowledge is indeed one way to "access" motivation, not just in the sense of what one knows, but also and more importantly in the sense of how one actively participates in knowing. In this way, to pay more attention to the visceral realm within social activism is also to recognize how knowledge is produced through social organizing, and who is included in the processes of knowledge production. As this research has shown, SGCPs sometimes rely on knowledge claims that preclude other ways of knowing, creating hierarchies between those who hold "correct" food knowledge (nutrition experts, teachers, food activists) and those who are by contrast naïve or lacking. Not only does this reinforce false and oppressive understandings of whose knowledge should count, but it also ultimately precludes alternative food activism from inviting the active, bodied participation of those that are "other" to the movement. The insistence on "correct," "pure" or "natural" ways of bodily knowing on the part of some SGCP leaders therefore hinders diversity and difference within alternative food. In order to attract participation from diverse groups, social activists need to begin to pay more attention to how the process of knowledge production within social activism can more effectively embrace the diversity of experience and ideas, practicing knowledge not as a static goal but as a collective process and progressive practice of situated, relational, and viscerally driven people.

Bodied education

Public education in North America has been impacted in recent years by cuts in federal spending, as well as increases in demands for standardization in testing (such as the Bush Administration's No Child Left Behind program). The effect of these conservative measures has been to severely restrict the concept of education itself. Many public schools have been forced to limit their educational curricula to "core" subjects like science and math, while dropping what are considered less important subjects of study, such as art and music. What counts as "legitimate" knowledge in our education system has therefore grown increasingly limited. More broadly, feminist scholar bell hooks (1994) notes that the increased conservatism of knowledge production in North American schools has also derived from the widespread conception of education that educator and philosopher Paulo Friere calls the "banking" system. In this system, knowledge is seen as something that teachers "have" and students "receive"; learning is not, in other words, a democratic, collective process.

This book has illustrated that attending to the role of the body within the processes of knowledge production is a valuable and constructive part of public education. The stories of the SGCPs that are presented in this book point to the need to broaden rather than limit the meaning of knowledge within public schools in the United States and Canada, particularly in terms that address the ways that different bodies come to know. The fact that so many schools have become concerned about food and nutrition suggests that such a broadening is possible, if not already underway, yet this broadening must go well beyond an expansion of accepted subject areas. As I have suggested, claims to "right" and "wrong" ways of knowing in schools can reinforce race, class, gender, and age-based hierarchies in the classroom, and can prevent students from being able to find agency within the educational system. As this work has shown, even as SGCPs are based upon innovative pedagogical strategies of hands-on learning, they can preclude active participation from certain individuals and groups by insisting on a "banking" system approach to alternative food. Nevertheless, this research has also illustrated that SGCPs can and do provide students with structured-yet-flexible spaces in which to negotiate their own roles in food decision-making. These lessons are important to the development of effective and inclusive food education in schools, but they also are significant beyond garden and cooking classrooms, even or especially within what have become the "core" subject areas of public education. Indeed, if education is to become relevant to a diversity of students and communities, it is important for teachers to attend to and encourage the active presence of minded-bodies within all aspects of formal learning.

In broadening what learning looks and feels like, and what counts as valuable knowledge, public education can become more responsive to all students' needs, and can provide important counter-experiences to the structured events of inequitable social life. Along these lines, education scholar Myra H. Strober (2003, 142) provides an important critique of the conservative trend in the education system, arguing that the ways in which education has been restricted and streamlined reflect a deep-seated neoclassical economic assumption that only market transactions produce value. The long-standing idea in the US that education can, and should, be responsive to the needs of society is certainly threatened by such a notion. As Strober points out, public education in the US has roots in the belief that education was central to the building of a better society, and to the functioning of a democratic political system. However, the individualism that is promoted by a neoclassical economist conception of education further limits these necessarily collective ideals (Strober 2003). Furthermore, hooks argues that

> the objectification of teacher within [such] bourgeois educational structures . . . denigrate[s] notions of wholeness . . . reinforcing the dualistic separation of public and private, [and] encouraging teachers and students to see no connection between life practices, habits of being, and the role of [teachers].
>
> (hooks 1994, 16)

In both of the above ways, the dominant conception of education in the US does not currently offer students a way to connect their classroom learning to their different life experiences, and does not offer them a way to make sense of or (re)negotiate their social position in light of their academic training. Rather, the current education system seems to reinforce the idea that learning has worth simply in the individualistic sense of market value. Yet this book suggests that the hands-on, interactive kitchen and garden classrooms of SGCPs have the potential to offer something different. Although SGCPs have, too, been critiqued as neoliberal (Pudup 2008), this research has shown that SGCPs can and do offer students opportunities to relate to new foods and food practices. SGCPs offer arenas in which students can "play" with their social identities vis-à-vis food. And, they offer models of collective work that do not (necessarily) promote individualism but rather suggest that the public school system is a place for the promotion of social health and welfare.

Both in ensuring active student engagement in knowledge production and in connecting education to broader social change, this SGCP research illustrates that it is important to recognize the dynamic role of bodies within educational settings. Although the body has, of course, always been present within formal education, its presence has not always been known or promoted as a means of active and meaningful engagement in the learning process. As education professor Stephanie Springgay insists, a truly bodied curriculum requires understanding and encouraging the presence of material bodies in the classroom not as docile and coded but as "redolent, fleshy, and becoming" (Springgay 2008, 72). This understanding points to how bodies are present in education particularly in unstructured ways, ways that preclude educators from being able to fully predict or control the path that learning takes, and that open the learning process up to the realm of the possible.

As this research has illustrated, it is important to recognize and allow for such unstructured, visceral work. In my own research methodologies, attention to the haphazard ontologies of SGCP events allowed me to recognize not only that SGCPs could both reproduce and resist the structures of social life, but also that both tasks could occur simultaneously. For example, as Chapter 8 discussed, pedagogical strategies like flexibility or repetition are involved in both such processes by simultaneously offering opportunities to viscerally access alternative food, and by discouraging or disabling such access. This recognition is important because it stresses that in attempting to construct a bodied curriculum, it is not possible to create blueprints for action. Lesson planning must instead be(come) a negotiated and collective process in which the intended outcomes cannot be fully known or preplanned, because the outcomes themselves are determined through collective engagement. Such a concept is inherently disruptive to the standardization of schooling that so many academics and educators have critiqued, and it is a way to shift public education away from a focus on the individual.

Towards feminist nutrition

In this book, I have also repeatedly expressed concern about the tactics and claims of nutrition science in promoting bodily health. These concerns have arisen in regard to the ways that knowledge is produced and policed, the eschewing of broader political economic contexts, and the lack of attention to the relational character of health/healthy experience. My study of SGCPs both confirms my worries about nutrition education as an exclusive and alienating project, and also points towards potentially helpful alternatives.

This SGCPs research has helped me to argue for a dismantling of expert-led knowledge within dietary science and nutrition education. In light of the contextual and situated ways in which students experience healthy, alternative food, it is impractical to think that food experts (either within or beyond the alternative food movement) could possibly come up with an answer to what everyone, everywhere, should eat. Instead, this research has shown that in order to encourage participation in any sort of nutrition program (such as SGCPs), it is important to acknowledge the presence and legitimacy of multiple forms of nutritional knowledge—both in cognitive and visceral terms. Indeed, my case studies suggest that it is necessary to highlight difference within nutrition education.

Other scholars have also articulated such a position. Geographer Gareth Enticott (2003), for example, wants us to become more attentive to the "lay epistemologies" (both in the realm of social life and in the wisdom of the body) that are important to nutrition and dietary choice. Thus, he suggests that we need to democratize science by extending it to lay knowledge's privileging of (bodily and communal) experience. In other words, nutrition education must begin from the understanding that there are multiple ways in which people have come to know food, and that all of these ways are legitimate and important in the advancement of health. By being attentive to such testimonies of experience, nutritional programs can challenge the limited, scientific understandings of healthiness, recognizing that there is not one nutrition but in fact many competing nutritions both within and beyond "accepted" nutritional science. In this scenario, (scientific) claims to expert knowledge are to be seen as unproductive, but the knowledge produced by dietary science is not to be totally rejected. Instead it is to be considered to be partial—existing alongside rather than above other ways of knowing (Haraway 1988).

Beyond dismantling expert knowledge, this research has illustrated that it is important to consider health as a situated and contextual phenomenon. This understanding recognizes the inequitable structure of health and health care in North America, as well as the broader unevenness of our social systems at large. In considering food preference not as an individual matter but as a visceral, socio-material production, I have argued that motivation to eat healthily, and indeed the visceral ability to "access" healthy food, must be examined in light of SGCP students' different social positions. Although SGCPs can hinder such an under-standing by promoting healthy eating as a universal act or a common experience, my fieldwork within Central and Plainville SGCPs shows that these programs also allow teachers and students to engage with food in more situated, contextual ways. Many SGCP teachers, for example, encourage students to experiment with recipes in ways that make sense to them, and also to experiment with other aspects of their lives, like racial or gender identities. In doing so, SGCPs stand to highlight that bodies are not just individual beings but relational processes that are necessarily in connection and negotiation with the various structures and systems that exist beyond the bounds of the skin.

Other scholars and health care workers have promoted a similarly broad understanding of health as well. Health scholars Robin Kearns and Wilbert Gesler, for example, endorse a similarly expansive definition of health as, "a state of complete physical, mental and social well-being and not merely the absence of disease or infirmity" (Kearns and Gesler 1998, 10). Likewise, geographer Andrew E. Collins suggests a notion of "sustainable health," a term that recognizes health as connected to other aspects of sustainable development, land and livelihood security (Collins 2001, 238). Public health policy scholar Meredeth Turshen offers Marx's definition: "A society in which men [sic], liberated from the "alienations" and "mediations" of capitalist society, would be the masters of their own destiny, through their understanding and control of both Nature and their own social relationships" (Turshen 1977, 59). And food policy scholar Timothy Lang argues in specific regard to nutrition concerns that, "nutritional science can and should contribute to social rather than individualized interventions" (2005, 731). All of these ideas are significant to SGCPs, but also more broadly signal the need for health care scholars and practitioners to increasingly take the situated contexts of social power and hierarchy into account. These discussions invite us to imagine how nutrition science and practice could become a politically progressive or even radical force in resisting oppressive and inequitable social structures.

Finally, my SGCP research highlights the need to consider nutrition in its broadest sense as a collectively produced and negotiated process. This understanding speaks to the relational ontology of nutrition—the ways in which nutrition knowledges and practices are produced through haphazard (yet also structured) networks of interrelating bodies, ideas, and things. In terms of SGCPs, this means that what emerges out of the programs is neither solely the product of the SGCPs themselves, nor clearly the result of one particular larger force (e.g. neoliberalism, fat-loathing, or racism). Such a recognition is important within nutrition education because it can allow students the opportunity to understand bodies not in terms of fixity and deviation from norms, but instead as both changing and changeable in relational connection to a variety of (powerful) social and biological forces.

In the pedagogical practices of SGCPs, as many teachers suggested to me, students often learn more about healthy food as a collective social process than they do about healthy food as a strict dictate of nutrition science. As my research has illustrated, SGCPs can provide space for students to learn how to find agency within the food systems both within and beyond the school. Although some students complained that strict rules or guidelines hindered their ability to take control of the kitchen activities, or of their eating habits more broadly, many students also suggested that SGCPs provided opportunities to relate to food and food ideas that were both helpful and fun. Moreover, SGCPs allowed students to interact with each other in myriad ways, through sharing stories or recipes to working and sitting together. As my research experiences illustrate, such opportunities are not tangential but rather vital to the success of SGCPs in motivating children to make healthy eating something that is (relationally) their "own."

Taken together these three shifts in nutrition education—from expert to partial knowledge, universal to situated health, and causal to haphazard explanations—amount to a new mode of "doing" healthy food that I have come to call "feminist nutrition" (Hayes-Conroy and Hayes-Conroy 2013). Much like the work of feminist scholars in promoting partial, situated, and collective knowledges, and the work of political ecologists in promoting contextualized understandings of health, feminist nutrition is based upon the understanding that any truly progressive nutrition program must emerge as a negotiated practice of particular people, in particular places and social positions. Feminist nutrition is a way to connect the socio-spatial understandings of human-environment geographers, and the politically inclusive work of feminist scholars, with the good intention and strong dedication of alternative and healthy food advocates. It is a framework that is specific to nutritional health, and yet it is also an example that could be relevant to many other health projects, as well as to the concerns of social activists and educators discussed above. In short, returning to the quotations that opened this concluding chapter, feminist nutrition can help to open the alternative food movement to the task of "listening in," and in doing so, can bring nutrition education closer to the elusive goal of "empowerment." As the title of this conclusion offers, it can urge all of us to recognize and respect the negotiated process that diverse food–body relationships necessarily take, where nothing is as simple as an either/or answer.

References

Ahmed, S. (2006) *Queer Phenomenology: Orientations, Objects, Others*, Duke University Press, Durham, NC.

Ahmed, S. (2010) *The Promise of Happiness*, Duke University Press, Durham, NC.

Alcoff, L. (2006) *Visible Identities: Race, Gender, and the Self*, Oxford University Press, Oxford.

Allen, P. and Sachs, C. (2013) "Women and food chains: The gendered politics of food," in P. Williams Forson and C. Counihan (Eds.) *Taking Food Public: Redefining Foodways in a Changing World*, Routledge, New York, pp23–40.

Anderson, B. (2005) "Practices of judgment and domestic geographies of affect," *Social and Cultural Geography*, vol. 6, no. 3, pp645–656.

Anderson, B. (2006) "Becoming and being hopeful: Towards a theory of affect," *Environment and Planning D*, vol. 24, no. 5, pp733–752.

Avakian, A. V. and Haber, B. (Eds.) (2005) *From Betty Crocker to Feminist Food Studies: Critical Perspectives on Women and Food*, Liverpool University Press, Liverpool.

Beard, J. (2007) *Beard on Food: The Best Recipes and Kitchen Wisdom from the Dean of American Cooking*, Bloomsbury Publishing, New York.

Beattie, J. (2002) *Women without Class: Girls, Race, and Identity*, University of California Press, Berkeley.

Bennett, K. (2004) "Emotionally intelligent research," *Area*, vol. 36, no. 4, pp414–22.

Berlant, L. G. (2011) *Cruel Optimism*, Duke University Press, Durham, NC.

Birke, L. (2000) *Feminism and the Biological Body*, Edinburgh University Press, Edinburgh.

Bondi, L. (2005) "Making connections and thinking through emotions: between geography and psychotherapy," *Transations of the Institute of British Geographers*, vol. 30, no. 4, pp433–448.

Bordo, S. (1993) *Unbearable Weight: Feminism, Western Culture, and the Body*, University of California Press, Berkeley.

Born, B. and Purcell, M. (2006) "Avoiding the local trap scale and food systems in planning research," *Journal of Planning Education and Research*, vol. 26, no. 2, pp195–207.

Bourdieu, P. (1984) *Distinction: A Social Critique of the Judgement of Taste*, Harvard University Press, Cambridge, MA.

Brennan, T. (2004) *The Transmission of Affect*, Cornell University Press, Ithaca, NY.

Brison, S. J. (2003) *Aftermath: Violence and the Remaking of a Self*, Princeton University Press, Princeton, NJ.

Brown, P. L. (2007) "A New Lease on Lunch," *The New York Times*, October 21.

Brown, P. L. (2008) "Bake Sales Fall Victim to Push for Healthier Foods," *The New York Times*, November 9.

Burros, M. (2009) "Obamas to Plant Organic Garden in White House," *The New York Times*, March 19.

Butler, J. (1990) *Gender Trouble: Feminism and the Subversion of Identity*, Routledge, New York.

Butler, J. (1993) *Bodies That Matter: On the Discursive Limits of "Sex,"* Psychology Press, London.

CA Gov (2005) "School Lunch," available at: www.cde.ca.gov/ls/nu/sn/nslp.asp, accessed January 2009.

Carson, R. (2002) *Silent Spring*, Houghton Mifflin Harcourt, Boston.

CBC (2006) "Junk Food," available at: www.cbc.ca/canada/nova-scotia/story/2006/09/12/junk-food.html?ref=rss, accessed December 2008.

Chen, M. Y. (2012) *Animacies: Biopolitics, Racial Mattering, and Queer Affect*, Duke University Press, Durham, NC.

Ching, B. and Creed, G. W. (1997) *Knowing your Place: Rural identity and Cultural Hierarchy*, Routledge, New York.

Clough, N. L. (2012) "Emotion at the center of radical politics: On the affective structures of rebellion and control," *Antipode*, vol. 44, no. 5, pp1667–1686.

Clough, P. T. and Halley, J. (Eds.) (2007) *The Affective Turn: Theorizing the Social*, Duke University Press, Durham, NC.

Collins, A. E. (2001) "Health ecology, land degradation and development," *Land Degradation & Development*, vol. 12, no. 3, pp237–250.

Colls, R. and Evans, B. (2008) "Embodying responsibility: Children's health and supermarket initiatives," *Environment and Planning A*, vol. 40, no. 3, pp615–631.

Corbett, M. J. (2007) *Learning to Leave: The Irony of Schooling in a Coastal Community*, Fernwood Publishing, Halifax, NS.

Counihan, C. (2008) "Mexicanas' public food sharing in Colorado's San Luis Valley," *Caderno Espaço Feminino*, vol. 19, no. 1, pp31–56.

Crang, M. (2002) "Qualitative methods: The new orthodoxy?," *Progress in Human Geography*, vol. 26, no. 5, pp647–655.

Crang, M. (2003) "Qualitative methods: Touchy, feely, look-see?," *Progress in Human Geography*, vol. 27, no. 4, pp494–504.

Crang, M. (2005) "Qualitative methods: There is nothing outside the text?," *Progress in Human Geography*, vol. 29, no. 2, pp225–233.

Cronon, W. (1995) *Uncommon Ground: Toward Reinventing Nature*, WW Norton & Co Inc., New York.

Davies, G. and Dwyer, C. (2007) "Qualitative methods: Are you enchanted or are you alienated?," *Progress in Human Geography*, vol. 31, no. 2, pp257–266.

Deleuze, G. and Guattari, F. (1987) *Thousand Plateaus: Capitalism and Schizophrenia* (Massumi, B. Trans.), University of Minnesota Press, Minneapolis.

Dewsbury, J. D. (2003) "Witnessing space: Knowledge without contemplation," *Environment and Planning A*, vol. 35, no. 11, pp1907–1932.

Domosh, M. (1997) "Geography and gender: The personal and the political," *Progress in Human Geography*, vol. 21, no. 1, pp81–87.

Dowler, L. and Sharp, J. (2001) "A feminist geopolitics?," *Space and Polity*, vol. 5, no. 3, pp165–176.

England, P. (2003) "Separative and soluble selves: Dichotomous thinking in economics," in M. Ferber and J. Nelson (Eds.) *Feminist Economics Today: Beyond Economic Man*, University of Chicago Press, Chicago, pp33–60.

Enticott, G. (2003) "Lay immunology, local foods and rural identity: Defending unpasteurised milk in England," *Sociologia Ruralis*, vol. 43, no. 3, pp257–270.

Gilligan, C. (2003) *The Birth of Pleasure: A New Map of Love*, Vintage, London.

Gregg, M. and Seigworth, G. J. (Eds.) (2010) *The Affect Theory Reader*, Duke University Press, Durham, NC.

Grosz, E. (1994) *Volatile Bodies: Toward a Corporeal Feminism*, Indiana University Press, Bloomington.

Guthman, J. (2004) *Agrarian Dreams: The Paradox of Organic Farming in California*, University of California Press, Berkeley.

Guthman, J. (2008a) "Bringing good food to others: Investigating the subjects of alternative food practice," *Cultural Geographies*, vol. 15, no. 4, pp431–447.

Guthman, J. (2008b) "'If they only knew': Color blindness and universalism in California alternative food institutions," *The Professional Geographer*, vol. 60, no. 3, pp387–397.

Guthman, J. (2012) "Doing justice to bodies? Reflections on food justice, race, and biology," *Antipode*, doi: 10.1111/j.1467-8330.2012.01017.x

Hall, K. Q. (Ed.) (2011) *Feminist Disability Studies*, Indiana University Press, Bloomington.

Haraway, D. (1988) "Situated knowledges: The science question in feminism and the privilege of partial perspective," *Feminist Studies*, vol. 14, no. 3, pp575–599.

Haraway, D. (1990) "A manifesto for cyborgs: Science, technology, and socialist feminism in the late twentieth century," in L. Nicholson (Ed.) *Feminism/Postmodernism*, Routledge, New York, pp190–233.

Hardt, M. and Negri, A. (2001) *Empire*, Harvard University Press, Cambridge, MA.

Hayes-Conroy, J. (2009) "Get control of yourselves! The body as ObamaNation," *Environment and Planning A*, vol. 41, no. 5, pp1020–1025.

Hayes-Conroy, J. (2010) "School gardens and 'Actually existing' neoliberalism," *Humboldt Journal of Social Relations*, vol. 33, no. 1, pp64–96.

Hayes-Conroy, A. and Hayes-Conroy, J. (2008) "Taking back taste: Feminism, food and visceral politics," *Gender, Place & Culture*, vol. 15, no. 5, pp461–473.

Hayes-Conroy, A. and Hayes-Conroy, J. (2010a) "Visceral difference: Variations in feeling (slow) food," *Environment and Planning A*, vol. 42, no. 12, pp2956–2971.

Hayes-Conroy, J. and Hayes-Conroy, A. (2010b) "Visceral geographies: Mattering, relating, and defying," *Geography Compass*, vol. 4, no. 9, pp1273–1283.

Hayes-Conroy, A. and Hayes-Conroy, J. (Eds.) (2013) *Doing Nutrition Differently: Critical Approaches to Diet and Dietary Intervention*, Ashgate Publishing, Surrey.

Hayes-Conroy, A. and Martin, D. G. (2010) "Mobilising bodies: Visceral identification in the Slow Food movement," *Transactions of the Institute of British Geographers*, vol. 35, no. 2, pp269–281.

Hinrichs, C. C. and Lyson, T. A. (Eds.) (2009) *Remaking the North American Food System: Strategies for Sustainability*, University of Nebraska Press, Lincoln.

Hochschild, A. and Machung, A. (2012) *The Second Shift: Working Families and the Revolution at Home*, Penguin, New York.

hooks, b. (1989) *Talking Back: Thinking Feminist, Thinking Black*, South End Press, Cambridge, MA.

hooks, b. (1994) *Teaching to Transgress: Education as the Practice of Freedom*, Routledge, New York.

Johnston, L. (1996) "Flexing femininity: Female body-builders refiguring 'the body'," *Gender, Place and Culture*, vol. 3, no. 3, pp327–340.

Johnston, L. and Longhurst, R. (2012) "Embodied geographies of food, belonging and hope in multicultural Hamilton, Aotearoa New Zealand," *Geoforum*, vol. 43, no. 2, pp325–331.

Kafer, A. (2013) *Feminist, Queer, Crip*, Indiana University Press, Bloomington.

Kearns, R. A. and Gesler, W. M. (1998) *Putting Health into Place: Landscape, Identity, and Well-Being*, Syracuse University Press, Syracuse, NY.

Kezar, A. (2003) "Transformational elite interviews: Principles and problems," *Qualitative Inquiry*, vol. 9, no. 3, pp395–415.

Kobayashi, A. (1994) "Coloring the field: Gender, 'race,' and the politics of fieldwork," *The Professional Geographer*, vol. 46, no. 1, pp73–80.

Kweli, T. (2007) *Eardrum*, Blacksmith Records, New York.

Lang, T. (2005) "Food control or food democracy? Re-engaging nutrition with society and the environment," *Public Health Nutrition*, vol. 8, no. 6A, pp703–737.

Lapine, M. C. (2007) *The Sneaky Chef: Simple Strategies for Hiding Healthy Foods in Kids' Favorite Meals*, Running Press, Philadelphia, PA.

Latham, A. (2003) "Research, performance, and doing human geography: Some reflections on the diary-photograph, diary-interview method," *Environment and Planning A*, vol. 35, no. 11, pp1993–2018.

Latham, A. and Conradson, D. (2003) "The possibilities of performance," *Environment and Planning A*, vol. 35, no. 11, pp1901–1906.

Latour, B. (1993) *We Have Never Been Modern*, Harvard University Press, Cambridge, MA.

Latour, B. (2004) "How to talk about the body? The normative dimension of science studies," *Body and Society*, vol. 10, no. 2–3, pp205–229.

Leitch, A. (2003) "Slow food and the politics of pork fat: Italian food and European identity," *Ethnos*, vol. 68, no. 4, pp437–462.

Livingstone, D. N. (1993) *The Geographical Tradition: Episodes in the History of a Contested Enterprise*, Blackwell Publishers, Boston.

Longhurst, R. (2001) "Geography and gender: looking back, looking forward," *Progress in Human Geography*, vol. 25, no. 4, pp641-648.

Longhurst, R., Ho, E., and Johnston, L. (2008) "Using 'the body' as an 'instrument of research': Kimch'i and pavlova," *Area*, vol. 40, no. 2, pp208–217.

Longhurst, R., Johnston, L., and Ho, E. (2009) "A visceral approach: Cooking 'at home' with migrant women in Hamilton, New Zealand," *Transactions of the Institute of British Geographers*, vol. 34, no. 3, pp333–345.

McCormack, D. (2003) "An event of geographical ethics in spaces of affect," *Transactions of the Institute of British Geographers*, vol. 28, no. 4, pp488–507.

McDowell, L. (1995) "Body work: Heterosexual gender performances in city workplaces," in D. Bell and G. Valentine (Eds.) *Mapping Desire: Geographies of Sexualities*, Psychology Press, London, pp67–87.

McIntosh, P. (1989) "White privilege: Unpacking the invisible knapsack," *Peace and Freedom*, vol. 49, no. 4, pp10–12.

McWhorter, L. (1999) *Bodies and Pleasures: Foucault and the Politics of Sexual Normalization*, Indiana University Press, Bloomington.

Massumi, B. (2002) *Parables for the Virtual: Movement, Affect, Sensation*, Duke University Press, Durham, NC.

Meatrix (2003) "The Meatrix," Free Range Studios and Sustainable Table, available at: www.themeatrix.com, accessed January 20, 2014.

Metzl, J. M. and Kirkland, A. (Eds.) (2010) *Against Health: How Health Became the New Morality*, NYU Press, New York.

Miranda, V. (2011) "Cooking, caring and volunteering: Unpaid work around the world," OECD Social, Employment and Migration Working Papers, No. 116, OECD Publishing. doi: 10.1787/5kghrjm8s142-en.

Mol, A. (2002) *The Body Multiple: Ontology in Medical Practice*, Duke University Press, Durham, NC.

Morris, J. L. and Zidenberg-Cherr, S. (2002) "Garden-enhanced nutrition curriculum improves fourth-grade school children's knowledge of nutrition and preferences for some vegetables," *Journal of the American Dietetic Association*, vol. 102, no. 1, pp91–93.

Morris, J. L., Briggs, M., and Zidenberg-Cherr, S. (2002) "Development and evaluation of a garden-enhanced nutrition education curriculum for elementary school children," *Journal of Child Nutrition Management*, vol. 26, no. 2, available at: www.docs.school nutrition.org/newsroom/jcnm/02fall/morris, accessed March 30, 2014.

Naples, N. A. (2003) *Feminism and Method: Ethnography, Discourse Analysis, and Activist Research*, Routledge, New York.

Nash, C. (2000) "Performativity in practice: Some recent work in cultural geography," *Progress in Human Geography*, vol. 24, no. 4, pp653–664.

Nestle, M. (2007) *Food Politics: How the Food Industry Influences Nutrition and Health*, University of California Press, Berkeley.

Nova Scotia (2006) "Fast Foods to be Phased Out of Schools Education," NS Health Promotion and Protection, September 12, www.gov.ns.ca/news/details.asp?id=2006 0912002, accessed January 20, 2009.

Orenstein, P. (2010) "The Femivore's Dilemma," *New York Times Magazine*, vol. 11.

Pérez-Rodrigo, C. and Aranceta, J. (2001) "School-based nutrition education: Lessons learned and new perspectives," *Public Health Nutrition*, vol. 4, no. 1a, pp131–139.

Pollan, M. (2003) "The (Agri)Cultural Contradictions of Obesity," *New York Times*, October 12.

Pratt, G. (2004) *Working Feminism*, Temple University Press, Philadelphia.

Probyn, E. (2000) *Carnal Appetites: Food Sex Identities*, Routledge, New York.

Probyn, E. (2005) *Blush: Faces of Shame*, University of Minnesota Press, Minneapolis.

Pudup, M. B. (2008) "It takes a garden: Cultivating citizen-subjects in organized garden projects," *Geoforum*, vol. 39, no. 3, pp1228–1240.

Raynolds, L. T. (2002) "Consumer/producer links in fair trade coffee networks," *Sociologia Ruralis*, vol. 42, no. 2, pp404–424.

Roberts, C. (2007) *Messengers of Sex: Hormones, Biomedicine and Feminism*, Cambridge University Press, Cambridge, MA.

Robinson, T. N., Borzekowski, D. L. G., Matheson, D. M., and Kraemer, H. C. (2007) "Effects of fast food branding on young children's taste preferences," *Archives of Pediatrics and Adolescent Medicine*, vol. 161, no. 8, pp792–797.

Rose, N. (2001) "The politics of life itself," *Theory, Culture and Society*, vol. 18, no. 6, pp1–30.

Rose, N. and Abi-Rached, J. M. (2013) *Neuro: The New Brain Sciences and the Management of the Mind*, Princeton University Press, Princeton, NJ.

Saldanha, A. (2005) "Trance and visibility at dawn: Racial dynamics in Goa's rave scene," *Social and Cultural Geography*, vol. 6, no. 5, pp707–721.

Sedgwick, E. K. and Frank, A. (2003) *Touching Feeling: Affect, Pedagogy, Performativity*, Duke University Press, Durham, NC.

Seinfeld, J. (2007) *Deceptively Delicious*, HarperCollins, New York.

Serano, J. (2009) *Whipping Girl: A Transsexual Woman on Sexism and the Scapegoating of Femininity*. Seal Press, Berkeley, CA.

Severson, K. (2008) "Slow Food Savors Its Big Moment," *The New York Times*, July 23.

Shiva, V. (2000) *Stolen Harvest: The Hijacking of the Global Food Supply*, Zed Books, London.

Slocum, R. (2007) "Whiteness, space and alternative food practice," *Geoforum*, vol. 38, no. 3, pp520–533.

Slocum, R. (2008) "Thinking race through corporeal feminist theory: Divisions and intimacies at the Minneapolis Farmers' Market," *Social & Cultural Geography*, vol. 9, no. 8, pp849–869.

Slocum, R., Shannon, J., Cadieux, K. V., and Beckman, M. (2011) "'Properly, with love, from scratch.' Jamie Oliver's food revolution," *Radical History Review*, no. 110, pp178–191.

Springgay, S. (2008) *Body Knowledge and Curriculum: Pedagogies of Touch in Youth and Visual Culture*, Peter Lang, New York.

Springgay, S. and Freedman, D. (2007) *Curriculum and the Cultural Body*, Peter Lang, New York.

Staeheli, L. A. (2001) "Of possibilities, probabilities and political geography," *Space and Polity*, vol. 5, no. 3, pp177–189.

Stewart, K. (2007) *Ordinary Affects*, Duke University Press, Durham, NC.

Strober, M. H. (2003) "The application of mainstream economics constructs to education: A feminist analysis," in M. Ferber and J. A. Nelson (Eds.) *Feminist Economics Today: Beyond Economic Man*, University of Chicago Press, Chicago, pp135–156.

Sundberg, J. (2004) "Identities in the making: Conservation, gender and race in the Maya Biosphere Reserve, Guatemala," *Gender, Place and Culture*, vol. 11, no. 1, pp43–66.

Taniguchi, H. (2006) "Men's and women's volunteering: Gender differences in the effects of employment and family characteristics," *Nonprofit and Voluntary Sector Quarterly*, vol. 35, no. 1, pp83–101.

Thien, D. (2005) "After or beyond feeling? A consideration of affect and emotion in geography," *Area*, vol. 37, no. 4, pp450–454.

Thompson, B. W. (1994) *A Hunger So Wide and So Deep: American Women Speak Out on Eating Problems*, University of Minnesota Press, Minneapolis.

Thrift, N. (2000) "Still life in nearly present time: The object of nature," *Body and Society*, vol. 6, no. 3–4, pp34–57.

Thrift, N. (2004) "Intensities of feeling: Towards a spatial politics of affect," *Geografiska Annaler: Series B, Human Geography*, vol. 86, no. 1, pp57–78.

Thrift, N. (2005) "From born to made: Technology, biology and space," *Transactions of the Institute of British Geographers*, vol. 30, no. 4, pp463–476.

Tsing, A. L. (1993) *In the Realm of the Diamond Queen: Marginality in an Out-of-the-way Place*, Princeton University Press, Princeton, NJ.

Turshen, M. (1977) "The political ecology of disease," *Review of Radical Political Economics*, vol. 9, no. 1, pp45–60.

Twine, F. W. (1996) "Brown skinned white girls: Class, culture and the construction of white identity in suburban communities," *Gender, Place and Culture*, vol. 3, no. 2, pp205–224.

USDA (2008) "Healthy Meals Resource System (HMRS)," available at: www.healthy meals.nal.usda.gov, accessed November 20, 2013.

USDA (2014) "USDA Farm to School Grant Program FAQs (Updated March 25)," available at: www.fns.usda.gov/sites/default/files/FY_15_F2S_FAQ.pdf, accessed April 2, 2014.

Valenti, J. (2007) *Full Frontal Feminism: A Young Woman's Guide to Why Feminism Matters*, Seal Press, Berkeley, CA.

Valentine, G. (2007) "Theorizing and researching intersectionality: A challenge for feminist geography," *The Professional Geographer*, vol. 59, no. 1, pp10–21.

Veugelers, P. J. and Fitzgerald, A. L. (2005) "Effectiveness of school programs in preventing childhood obesity: A multilevel comparison," *American Journal of Public Health*, vol. 95, no. 3, pp432–435.

Vocks, S., Stahn, C., Loenser, K., and Legenbauer, T. (2009) "Eating and body image disturbances in male-to-female and female-to-male transsexuals," *Archives of Sexual Behavior*, vol. 38, no. 3, pp364–377.

Walkowitz, D. J. (1999) *Working with Class: Social Workers and the Politics of Middle-Class Identity*, UNC Press Books, Chapel Hill, NC.

Waters, A. (2007a) *The Art of Simple Food: Notes, Lessons, and Recipes from a Delicious Revolution*, Clarkson Potter, New York.

Waters, A. (2007b) "Slow food, slow schools: Transforming education through a school lunch curriculum," available at: www.present-tense-living.com/raised_bed_kitchen_gardens/pdf_files/slow_food_in_schools.pdf, accessed April 3, 2014.

Waters, A. (2009) *Edible Schoolyard: A Universal Idea*, Chronicle Books, San Francisco.

Waters, A. (Online, n.d.) "The Delicious Revolution," Center for Ecoliteracy Publications, available at: www.ecoliteracy.org/publications/rsl/alice-waters.html, accessed January 20, 2014.

Waters, A. and Berkeley, C. (2006) "Eating for Credit," *The New York Times*, February 24.

Whatmore, S. (2002) *Hybrid Geographies: Natures Cultures Spaces*, Sage, New York.

Williams, P. (2002) "Thought About Food, A Workbook on Food Security and Influencing Policy," Nova Scotia: Atlantic Health Promotion Research Center, available at: www.foodthoughtful.ca, accessed April 3, 2014.

Williams, P., Mount Saint Vincent University, Health Promotion and Protection (2007) "Cost and Affordability of a Nutritious Diet in Nova Scotia," Report of 2007 Food Costing, available at: www.novascotia.ca/dhw/healthy-communities/documents/Cost-and-Affordability-of-a-Nutritious-Diet-in-Nova-Scotia-Report-of-2007-Food-Costing.pdf, accessed April 3, 2014.

Williams, P. L., Johnson, C. S. J., Johnson, C. P., Anderson, B. J., Kratzmann, M. L. V., and Chenhall, C. (2006) "Can households earning minimum wage in Nova Scotia afford a nutritious diet?," *Canadian Journal of Public Health*, vol. 97, no. 6, pp430–436.

Williams-Forson, P. and Counihan, C. (Eds.) (2013) *Taking Food Public: Redefining Foodways in a Changing World*, Routledge, New York.

Zimmerman, H. (2013) "Mobilizing caring citizenship and Jamie Oliver's food revolution," in A. Hayes-Conroy and J. Hayes-Conroy (Eds.) *Doing Nutrition Differently: Critical Approaches to Diet and Dietary Disease*, Ashgate, Surrey, pp277–296.

Index